A Celebration of a Century

A CELEBRATION OF A Century
ONE HUNDRED YEARS OF FAMILY HARMONY

SYLVIA ENGEN

Belleville, Ontario, Canada

A Celebration of a Century
Copyright © 2004, Sylvia Engen

All Rights Reserved. No part of this publication may be reproduced, stored in a retrieval system or transmitted in any form or by any means—electronic, mechanical, photocopy, recording or any other— except for brief quotations in printed reviews, without the prior permission of the author.

Scripture quoataions marked NRSV are from the *New Revised Standard Version* of the Bible, copyright 1989, by the Division of Christian Education of the National Council of the Churches of Christ in the United States of America, and are used by permission. All rights reserved.

ISBN: 1-55306-852-1

For more information or to order additional copies, please contact:

15 Montjoy Avenue
Camrose AB T4V 2L3
Tulsa, OK 74136 USA

Guardian Books is an imprint of *Essence Publishing,* a Christian Book Publisher dedicated to furthering the work of Christ through the written word. For more information, contact:
20 Hanna Court, Belleville, Ontario, Canada K8P 5J2.
Phone: 1-800-238-6376 • Fax: (613) 962-3055.
E-mail: publishing@essencegroup.com
Internet: www.essencegroup.com

Printed in Canada
by
Guardian BOOKS

I am appreciative to my family for encouraging me to write what I have heard and gleaned. I am appreciative to Edith Dayton for proof reading the manuscript and to Solveig Johansen and Ellen Auckland for translating the Norwegian letters.

Table of Contents

Introduction

1. The Immigrants. 13
2. The Homesteaders 23
3. Mary Pederson 33
4. Charlie Pederson. 41
5. Facing the Fear of Nature. 45
6. The Worn Black Purse 49
7. The Red Velvet Album 53
8. Erwin Henry Pederson 59
9. Ida Josephine Pederson Hagen. 63
10. Clarence Selma Pederson 67
11. Arnold Glen Pederson 71
12. Mrs. Arnold Glen Pederson 77
13. The Turbulent Thirties. 83
14. Maple View School Days 87

15. The Stately Barn on the Pederson Place....... 91
16. Family Gatherings........................ 95
17. Ole, the Homestead Neighbour and Cousin ... 99
18. Introducing the Forties................... 105
19. Ice for the Ice House 113
20. The Aladdin Windsor 117
21. A Visit from Brother Andres............... 121
22. The Fifties and Sixties.................... 125
23. The Close of a Century................... 137

How Do I Record a Celebration of a Century?

The year is 2004. Our family farm has been "the Pederson Place" for one hundred years. We anticipate a celebration, a gathering together of our clan, friends and neighbours to pay homage to our homesteader grandparents who came, planted and toiled on this sacred soil. The family requests that the farm memoirs be put into print.

How do I begin to record as I reminisce? Where do I start, as I attempt to put the history of our farm in permanent form? I concur with the Psalmist, *"I have a goodly heritage"* (Psalm 16:6b NRSV). But my ancestors, who could give me facts, share their experiences and answer my many questions, are no longer living. I can only write what I know, gleanings from old letters, pictures, the community history book and stories passed down in response to our frequent plea, "Tell us about the olden days!"

I am overwhelmed with the task of compiling a farm history that covers a hundred-year period, from the time

~ *A Celebration of a Century* ~

my grandparents started the homestead until today. After thoughtfully contemplating my options, I know that I have to physically return to the home of my birth. Nostalgic memories tug at my heart; I want to feel and record the emotions that are part of me.

Pictures stored in my memory flash before me as, pen and notepad in hand, I set foot on the soil of the farm that was my home in my growing-up years. I feel the gnarled walls of the old granary built by Grandpa Charlie. The knotted and parched pores of each slab of wood give evidence of blistering sunshine, high winds and blustering blizzards. The bulging boards speak of faithful endurance, as they contained bushels of grain year after year. *Oh, aged granary, why can't you speak to me?*

I stumble through untamed grass, skirt a slough and climb over chunks of broken cement as I make my way to what had been Grandma Mary's junk pile. It had provided a collection of treasures that were hauled into the trees where they furnished playhouse after playhouse in the years of my youth. Now I recognize a rusted reservoir of the old McClary cook stove, pieces of a crock that saw my mother's miracle cucumbers become sweet pickles and fragments of an aged cutter yoke that manipulated Queen, our faithful horse, as we drove to school in winter months. *I'm transferred to days of long ago as I step through your mound of memorabilia, Old Junk Pile! I yearn to hear the stories hidden in the depths of your debris!*

Overcome with nostalgia, I continue to explore the remnants of our farm, determined to continue recording each thought and feeling that flood my being. An old treadle sickle sharpener, tucked in the trees, catches my attention. A vision of Dad, sitting on the seat attached to the wooden frame that upheld two cement grindstones, flashes before me. I can hear the rhythmic swish of water from the trough as Dad turns the handle to sharpen and smooth each sickle before haying season. I remember begging as a child, "Let me turn the handle, too!" I reverently touch the rough, worn surface of the grindstone as I question, *Did Dad build you, or are you Grandpa Charlie's creation?* Only a robin breaks the silence, with a soothing song.

My steps continue to take me past varied memorabilia, one after another. But wait! I am going to elaborate further on their significance within the chapters of this book…I realize that my recollections of the farm itself are only the beginning of my research. They spark my enthusiasm to search, explore further and question any memory, story or document that might pertain to our family farm.

My tour has evoked stronger memories and brought increased harmony with my heritage to the core of my being, to my attachments and aspirations. It has evoked and magnified sentiments about my roots. The legacy left me is more firmly embroidered in my heart. I trust my farm history will convey the wonder and appreciation of our forefathers that now overwhelms me. Our kin can be proud that, one hundred years ago, Grandpa Charlie and Grandma Mary faced many challenges with faith and fortitude as they paved the way, establishing a farm home for themselves and their descendants.

I can only write from the heart as I record. We may all differ as to the recollections we each cherish from the past, but we have to agree that we are blessed with a rich heritage!

~ *A Celebration of a Century* ~

1904
The Charles Pederson Family
Left to right: Erwin, Charles, Mary and Ida
Seated: Clarence and Arnold

The Immigrants

Life in Appelton, Minnesota, in 1904 was not easy for Mary Pederson. In the 1800s, Norwegian immigrants—including Mary—had flocked to America, especially to the upper midwestern states. Norway's population was growing. Young people had little opportunity to make a living. Mary, with Mother Ingeborg and seven siblings, the youngest three years of age, braved the long sea journey and landed in New York in 1887. They then travelled by train to Appleton, Minnesota.

Mary and her family came from the Haugen farm on the steep slopes of a mountain in Hallingdal. The barren terrain, with only thin layers of soil, left too few tillable acres to yield sufficient produce for a family of fifteen. Mary knew hardship. The rugged beauty of the land that gave her birth also brought pain. The rocky land was hard to run on and cold in the winter when there were no shoes to wear.

Spring always brought hope and relief. The cows and sheep were starved after spending the long winter months in the crowded stalls of the Haugen barn built against the mountain. The barn was cleared of the accumulated waste and the fertilizer spread on the fields. Mary helped drive the animals, their bones protruding, up the mountain to the fertile grass areas near the *seter*, a stone-and-wood shelter, for the summer months. The dirt-floored *seter* provided some protection, but streams flowed near the animals' bedding during rainstorms. (Mary was to learn later that her older sister, who remained in Norway, gave birth to her son in that *seter*.)

Mary tended the animals as they grazed on the park-like mountainside. She was never allowed to forget the morning a cow stepped on a bare foot and left a lifelong scar. Days could be long for thirteen-year-old Mary up in the *seter*, but when butter and cheese making were completed each day, the magnificent beauty of the wildflowers and lush green growth of the mountaintop compensated for the loneliness. She never complained as she shared the family responsibilities. Mary exemplified the dependable, positive and steady characteristics of a typical Norwegian who lived close to the land. The way of life was primitive. Their livelihood was a meagre one. But the Haugens were a free and courageous family, true to the nature of the Norwegian people.

A story is told of carpenters who built a building on the farm, using a handsaw to cleave the lumber. It was winter and extremely cold. Many lacked warm apparel. One of the carpenters wore pants, made of animal skins that were more or less in rags. He worked faithfully, with skin exposed to the cold elements because he never owned underwear. Do not all Norwegians persevere if there is work to be done?

~ The Immigrants ~

The roughly built log schoolhouse, halfway down the mountain, opened the doors of learning for Mary. Education was important to the Haugens. Mary's grandfather was a descendent of the Soops, who were of noble blood. The most notable was Archbishop Olav Engelbriktson of Nidrraros (Trjondheim). Olav, who attended the University of Rostock, was president of the Norwegian Council in the sixteenth century. He built himself a stone fortress on the Trondheim fjord and accumulated soldiers and ships. The Soops' coat of arms depicted three yellow lilies in a red pitcher of clay. This coat of arms can still be seen in Aurland Church, Sognefjord, on the pulpit from the thirteenth century. The Soops owned the church for several decades.[1]

Norway, rising furrowed and weather-worn from the sea, was a land of many mountains and valleys, with a sea to the north, west and south. It was a great expanse of barren rock, ice fields, dense forest, mountain lakes, swamps, chasm-like valleys and numerous fjords—lengthy indentations of the sea reaching inland from the coast. The ocean posed no obstacle to the Norwegian people. It was the gateway to lands of prosperity. There was little room for opportunity for the young; hence, when news came that there was land available across the ocean, men and women of pioneer vision, lured by the call of a brighter future, left their homes for a better tomorrow.

The Norwegian peasant had never known bondage. His lot had been difficult, but he had never ceased to be a free man. Mary's father was a true Norwegian individual, fiercely resisting any encroachment upon his personal freedom. To hard-working Knut Haugen, struggling to keep above poverty on a small plot of tillable land, reports of ample land in America came at an opportune time. Mary's father knew it took the whole of man's waking hours to win a livelihood from the soil of the hillside farm.

And so it was decided that Mother Ingeborg would sail to America. Father Knut was too ill to travel. Preparations to leave brought Mary pain. Norway, with its majestic beauty, was the country she loved. *"Ja, vi elsker dette landet"* ("Yes, We Love this Country"), the national anthem of Norway, expressed the heartfelt sentiments of the people for their beloved land.

Mary was the eldest of the eight children who took the long journey by sea to America in 1887. Seasickness, homesickness and crowded quarters were minor factors in the lengthy and difficult trip. They lived through the journey, although there were many reports of ships that ran aground on dangerous reefs and rocks, ships damaged by fire, outbreaks of illness, low food supplies and passenger deaths. To Mary's family and to all the Norwegian people, the oceans and seas did not pose a problem; they were their highways to new horizons.

Kari, Mary's older sister, had married in Norway so did not emigrate. Mary had been her faithful helper. Mary told her own family in later years that Kari wept for children of her own. She and her husband Knut had been married for eight years. Later they had eight children. She wept again because she had too many.

Mary remembered the day a childless neighbour was asked to visit Kari and Knut's home. "The parents offered to give a child away. They went from bed to bed but found out that they could not part with any of their children."

Kari knit Norwegian mittens; the mittens became popular in the area. Soon she was knitting for royalty. The

~ *The Immigrants* ~

family raised sheep. They sheared, cleaned and dried the wool. The wool was carded into small strips and spun into skeins. A meagre livelihood was supplemented by their mitten production. Kari Brye's sons were photographers. Today her grandsons are professional photographers.

Far from the shores of Norway, a distraught, hungry family, unable to speak English, landed in New York. It was the Seamen's Mission that fed the family and put them on a train for Appleton, Minnesota. Immigrants who had left Hallingdal earlier were awaiting the arrival of Ingeborg and her eight children. One such immigrant was Halvor Robertson, a young, hard-working and prosperous farmer. He built a small house on his land for the Haugens. It stood among rolling hills that were dotted with stones to remind them of their home in Norway. A stable large enough for a horse and a cow was erected. The support, compassion and caring pioneer spirit abounded among the immigrants.

Father Knut, recovered from a severe illness, and the older siblings sailed for America in 1889 to join the family in Minnesota. Sigrid, the eldest daughter, and her four young children also arrived with her father. Her husband, twenty-nine years of age, had died in 1885. She accepted the position of housekeeper for the kind bachelor, Halvor Robertson. They were married in 1890. Four little children again had a father. The youngest of Sigrid's children, Ole H. Olson, who was four years old when he arrived in Minnesota, built a home adjacent to the Saskatchewan Pederson homestead in 1912.

Mary again experienced loss when, shortly after the family was reunited, brother Lars drowned in Lac Qui Parle Lake, which bordered their new home. The new immigrants struggled to produce vegetables on virgin soil and sold them to the Appleton Hotel. Father Knut transported the produce in a small one-horse wagon. They lived close to the earth, and their vitality and ingenuity patched together a living.

The Haugens soon became members of Lac Qui Parle Lutheran Church, which had been organized in 1870. The pioneer women had formed a Ladies' Aid group. God-fearing men and women welcomed the new family. The church was the hub of the community, just as the *Hol Stav Kirke* (Hol Stav Church), a preserved church of the twelfth century, was the centre of the community back in Hallingdal, Norway. Mary and her fourteen siblings had been baptized and regularly worshipped within the walls of their sacred church. Mary, born May 21, 1874, was baptized there on June 14, 1874. She was given the name "Marita Sletto" after her father's first farm. The church, built in 1192, stands today—a monument to the many who have worshipped at *Hol Kirke*.

The Haugens were typical Norwegians, resilient people who displayed a timeless strength, a sense of immortality that embraced the capacity to move forward despite hardship and repression. Inspired by their faith in God, hard work prevailed, and Mary's family prospered well until the 1890s when choice land ran out. Just as in Norway, there wasn't ample soil to till. As the settlers of Norwegian blood continued to pour westward across the Mississippi, available land was difficult to find.

Numerous new immigrants claimed the name of *Haugen*. So Father Knut Haugen changed the family name, as the custom had been in Norway, to *Knutson*. He reasoned, too, that Haugen referred to a hill in the Norwegian language. The Haugen farm in Norway, built on a high hill, was left behind. Knut's offspring would be known as Knutsons.

Ultimately there was no opportunity to homestead additional land in Minnesota. More and more young people wanted homes of their own. Food was sparse for the Knutsons and their new neighbours. The Norwegian families, who consistently withstood chaos, were challenged.

The 1890s were eventful years for Mary Knutson. She met and married Charles S. Pederson, son of Sebjorn and Gjertrud (Kittilsen Sebjornsdatter) Pederson. His parents and two oldest brothers had immigrated from Lisledal, Norway, in 1861. His eldest brother, Peder, was born in Norway in 1858. (Peder's eldest child, Dr. Nellie Holman, was a doctor and a missionary to China from 1919 to 1931. Another daughter, Gertrude M. Hartman Hanson, was the regional vice president of the League of Minnesota Poets from 1935 to 1950. She was poetry editor of *Friend* magazine and did editorial work on several books. Her legacy of poetry was published in the book *Come Walk With Me*, and she is especially remembered for her poem, "God, Light Tall Candles in My Heart.") Charlie's brother Sebjorn (Sam) was born in 1860. He was one year old when he arrived in America with his parents. (Sam and his wife lost five babies in infancy. One of his granddaughters, Arlette Villaume, was president of the Daughters of the Reformation for several years.) Charlie was born in Minnesota on December 24, 1868. Two sisters, Annie and Gro (Grace) were born later.

Charlie and Mary's wedding took place in Lac Qui Parle Church. A three-room house, across the road from the original Pederson homestead in Hantho township of Lac Qui Parle county, became their home. Four children—Erwin, Ida, Clarence and Arnold—were eventually added to the family. Charlie and Mary realized there wouldn't be work opportunities in the area for their growing family, and they began to search for relocation possibilities. Mary knew the sadness of saying farewell. But she was a stalwart Norwegian, characterized by warmth and courage. She quietly faced any problem, no matter the dimensions.

Notes

[1] Kristian T. Knutson, *Kleppo Sleksbok* (Norway: Eget Forlag Publisher, 1972) p. 123.

~ A Celebration of a Century ~

1903

*A Sunday gathering outside of Charles and Mary Pederson's sod homestead dwelling.
Mary is wearing a white blouse, a light colored skirt with a black belt and stands in the middle of the picture.
Son Arnold, wearing a white shirt and dark trousers, stands holding his father's hand to the left of the picture.*

The Homesteaders

Charlie Pederson heeded the glowing accounts, given by Canadian immigration agents, of fertile land available for anyone with a pioneering spirit. He was told that the Canadian Pacific Railway had been given a land grant of all the odd-numbered sections within a distance of twenty miles on both sides of the main line running from Portal to Moose Jaw in Saskatchewan. He was also told that men over the age of eighteen could purchase a quarter section of land for ten dollars. The purchaser was required to erect a dwelling on the land, reside there for at least six months of the year for a period of three years, and break at least fifteen acres of land during that three-year span. This was termed "proving up" a homestead. At the end of the three-year period, if all the requirements were fulfilled, they received free title to the land and became naturalized Canadian citizens.

Charlie must have been intrigued by the promise of available land. In 1902, he went across the Canadian border with a scouting party to investigate homestead possibilities and ventured twelve miles into Saskatchewan. A statutory declaration signed by Charles S. Pederson, dated July 17, 1905, found in the Saskatchewan Archives in Regina, stated that the land breaking required was completed on the "N.E.32-2-11-W2" by October 2, 1902. The Cambria municipality history book from Torquay, Saskatchewan, reports that "Charles S. Pederson, on October 20, 1902," signed a petition that requested a "re-survey be made" of the land boundaries that he had "filed on." Charlie's statutory declaration form stated that he erected a house, fourteen feet by sixteen feet, at a cost of eighty dollars, in June of 1903. It also confirmed that Charlie lived in this house in June and July of 1903.[1]

And so it was in 1904 that Charlie Pederson moved his family from the Appleton area and set out for Saskatchewan. The trip could only be taken on the Soo Line Railway as far as Flaxton, North Dakota. The railway from Flaxton to Ambrose was not built until 1906. The family of four, with only the necessities of life, made their way across the Canadian border and twelve miles north into Saskatchewan to settle on their homestead near Turner Post Office, a small postal station in a neighbour's house. The date, according to the statutory declaration, was February 27, 1904.

This land was their home, but it was a lonely sod house set in a desolate sea of white snow that barely covered the previous year's tall, protruding grass and naked brush. It was a vast sea, going on for miles and miles, broken only by protruding mounds of stones that dotted the skyline. It was cold, and the sod house had no insulation. Eldest son, Irwin, was ten years old; Ida was six;

~ The Homesteaders ~

Clarence was five; and Arnold, the youngest, was two. Dreams of a new beginning with abundant opportunities for a good life must have been shattered for Mary and Charlie during their first winter of 1904. But they persevered. They put their trust in God.

Signs of spring no doubt cheered the hearts of the family of six who had been confined to the small sod house. But the burst of new waving grass continued into the skyline, as did the snow of winter. The few acres of newly broken land emerged, rock studded and soil upheaved, as the snow disappeared. The challenge to the sod house dwellers was immense and daunting.

A picture in Mary's album shows the sod home in 1906. The picture must have been taken on the Sabbath, as the Pedersons and their neighbours wore Sunday clothes, perhaps the only good clothes they possessed. One can presume that the homesteaders had gathered on the Pederson farm for a church service that day. The barn built by Charlie with the help of son Erwin is shown in a 1908 photograph of a community gathering. Neighbours helped in the construction of the building, which served for years as a haven for horses, cows, pigs, cats and hens.

Homesteaders reported that the first years after the turn of the century were known as "the rainy years." It was difficult to live under a sod roof when there was excessive rain for seven years. Mary later told her family that the only place she could keep her flour dry was under the table. The Pederson homestead is located a distance from the main road. No doubt the farm was established on the only dry spot Charlie could find.

It has been said that mud worked well as an outside-wall plaster and white wash gave a finish to the inside walls of the sod house. Mary likely used newspaper to paper a portion of the walls in a decorative fashion. The furniture was simple and homemade. A stove was a necessity. Wooden packing boxes were turned into cupboards, tables and chairs. Homemade furniture, they say, was not held together with nails but with wooden pegs. Newspapers cut into patterns with fringed edges were another of Mary's creations to take the place of curtains for the small windows.

It isn't difficult to imagine Mary's ingenuity in making quilts for Charlie's homemade log beds. She also made mattress covers from whatever material she had and stuffed them with straw. This task took place after the first harvest, when the straw was fresh and clean. Erwin, Ida, Clarence and Arnold had great fun climbing on the high, rounded mattresses. Before long the curved tops were flattened considerably as the children used their beds.

Homesteaders remember the unwelcome visitors that also occupied the beds. Bedbugs resembled small ladybugs and came with the logs and lumber used in building. The bed frames were usually wooden, and cracks made ideal places for the bugs to hide. These creatures feasted on human blood at night, disturbing their victims' sleep. They could be tricked into emerging by blowing out the coal oil lamp, waiting for a half an hour, lighting the lamp again— and quickly killing any bold bed bug. Before beds were made each morning, a thorough hunt was made to capture any wee molester that came into view.

Hot summer months were not conducive to a good night's sleep. Mary was known to dip sheets in water and then hang them around the beds to help bring a sense of coolness to distraught would-be sleepers. Beds were often moved to the granary in summer months, as it promised a breeze and fewer bedbugs.

Charlie worked diligently to prepare the land for maximum grain yield. He had broken the small plot of land with his four horses and a wooden beam plow. The earliest tillage equipment consisted of an axe and a single-blade walking plow. Brush was cut with a scythe and piled before the land could be broken. This was hard, back-breaking work. Homesteaders helped by loaning animals to neighbours who didn't own oxen or horses but were desperate to break a few acres. A story was told of a homesteader who spent hours by a slough on a hot day, waiting for his oxen to leave the water. They wouldn't budge to his threats until they were ready to resume their task.

Inadequate, primitive farming equipment, frost, smut germs and unsuitable varieties of wheat made it difficult to secure an adequate supply of flour, so essential for daily food needs. Harvesting and thrashing of grain presented problems to the early homesteaders. A scythe with a cradle attachment was often used for the first-harvest crop.

No doubt there were Norwegian settlers who used the flail to harvest their grain, as was done in Norway in the eighteenth century. Bundles of grain were laid in a circle, heads toward the centre, on a small area of ice. A man with a flail beat the grain. The straw was cleared away with a fork, but the chaff remained. The grain was tossed for short distance into the granary, clearing the grain of chaff. The largest and best kernels fell inside the building, while the lighter kernels fell outside. The wind blew the chaff away. The grain was run through a fanning mill and ground into flour.

There were hardships, but nature attempted to compensate with bountiful blooms; wildflowers thrived on the open prairie. Crocuses, buttercups and buffalo beans decorated the landscape and Mary's table. Silver willows added their beauty. Prickly bushes dotted with pink roses were everywhere. The striking bloom of the cactus brought delight to the eye but agony to a bare foot. Tiger lilies were found, as were small daisies. Autumn was glorified with the golden rod and bright fireweed. The dandelion is one of the few flowers that have endured a century of blooming on the Pederson farm.

Fortunately for the Pedersons, the first winter was mild. However, the following winters were bitterly cold. The sod house was primitive. Blankets would often be frosted by morning. Water containers would freeze if the fire went out in the night. Stories are told of gathering cow chips to use as fuel in the stoves. Wood wasn't abundant on the open prairie, but cow chips were always available. Isolation in the vastness of the prairies in winter, when nature could be merciless, brought loneliness. We can understand why bachelors in the area moved in together.

There were no proper roads, only trails, sod paths or graded dirt that became quagmires after a rain. No ditches meant no runoff when water puddles became a slough.

Tracks eventually disappeared in the water. Stories were told of homesteaders who drove for miles in water that reached the hubs of wagon wheels during the heavy rains in 1904.

It was a challenge for Charlie to keep track of his cow and horses on the wide open spaces of his homestead where the sky stretched blue in four directions and the land was virtually flat and treeless under it. Animals were often equipped with bells so they could be located if they wandered out of sight. Herding the cattle was a necessity when fences were not yet in place. Stock roamed as they pleased. Children grew up quickly as they assumed responsibilities, such as herding the cows each day.

It was the custom of the women in Norway to faithfully care for the animals when the men were working in the fields. The women often did the milking. Hence, Mary was accustomed to the task of milking, the task that became hers on the Pederson homestead. Erwin, as helpful oldest son, worked at her side. Mary felt at home with the horses as well as with the cows. She was Charlie's assistant during haying as well as in the seeding and harvesting of the annual grain crop.

The Pedersons were homesteaders who courageously helped lay the foundation for their community. The first church services were held in the homes. It was important for them to have a congregation established. The Lac Qui Parle congregation was organized on March 27, 1905, according to the *Cambria History Book* published in 1978. Lac Qui Parle Scandinavian Lutheran congregation was named after the Lac Qui Parle Church, the beloved church left behind in Minnesota.

People met in their homes to worship. Walking was the Pedersons' mode of travel to the neighbours when weather permitted. A Ladies' Aid group was formed. The members were diligent in gathering finances to pay a pastor's salary and prepare to build a church. The *Cambria History Book* goes on to say that at the July 4th meeting of the new congregation in 1905, it was decided to pay $200 to Home Missions toward the support of a pastor. Imagine the jubilation of the community when sufficient monies were collected to make plans for the construction of a church edifice in 1912. The Lac Qui Parle church became a reality in 1915.[2]

~ A Celebration of a Century ~

A green velvet-covered foot warmer upstairs in the granary was a reminder of winter trips to church with horses and a wagon. A metal container in the foot warmer would be filled with hot ashes. Stories are told of a constant mosquito invasion during the warmer months. Mary handled the problem by covering the children's heads with gunnysacks as they travelled to church.

Education was important to the homesteaders. The Department of Education was located in Regina, Northwest Territories, in the early days. In order to have a school district formed, it was necessary to forward information about the proposed location, a name and how it would be financed. The site had to be in the centre of the proposed school district. Money had to be raised to build and equip the schoolhouse. A grant from the Department of Education, based on attendance, would ensure the school would continue once it was constructed. The Maple View School district was formed in 1907. The *Cambria History Book* reports that Charlie Pederson was one of the first school trustees. Classes were held in a home in 1907. The school was built on North East 33-2-11 in 1908, the year young Arnold started school. Contractors were paid $325. The contract, as quoted in the *Cambria History Book*, stated," "In case of the work being done on or before October first, the compensation to be $350."[3] Additional bills show that five pounds of nails cost twenty-five cents, students' double desks cost five dollars each, a teacher's desk cost fifteen dollars, a stove cost twenty dollars and a broom was worth forty cents. [3]

The school provided numerous opportunities for fellowship. Church services and Sunday school previously held in homes moved to the school until a church was built. The community now had a gathering place. The school gave the Pederson children a place to have fun. Games, such as hide-and-seek, anti-I-over, and stealing sticks, were enjoyed at recess. The picnics, ball games, Christmas concerts and evening socials at the school were highlights for the community.

The homesteaders enjoyed the community functions as a respite from their farm duties. One such duty was to continue clearing land. Charlie and his sons spent hours picking rocks; rock piles presently on the Pederson farm confirm their aged existence despite a century of wind and rain. There were no shortcuts to clearing brush and rocks from untamed soil. The chainsaw hadn't been invented in 1904. It took an axe and strong backs for the Pedersons to prepare the land to grow adequate grain and produce to meet their daily needs.

Notes

[1] Fifty and Over Club, *Our Prairie Heritage* (Altona, Manitoba: Friesen Printers, 1978) pp.14-15.

[2] Fifty and Over Club, p. 98.

[3] Fifty and Over Club, pp. 118-119.

~ *A Celebration of a Century* ~

1908
The Charles and Mary's Homestead
The Pederson's stand to the right of the picture. Erwin, with the bike,
stands to the extreme right, then Ida, her mother Mary,
her father Charles and Clarence.
Arnold, clutching his dog, stands in front of his father.

Neighbours with their transportation parked near the barn,
have come for a visit.

Mary Pederson

The Sunday-company pillowcases, white as snow and ironed to perfection, intrigued Mary's grandchildren and her visitors. An embroidered message immaculately stitched with satin thread read, "I slept and found life was beauty; I awakened and found life was duty." Their owner treasured the pillowcases, only brought out for special occasions. Her grandchildren's speculations were varied, but only Mary knew the meaning of her pillow's message.

The capable, courageous and devoted Mrs. Charlie S. Pederson faced many lonely hours being "dutiful" on the primitive homestead. The neighbours were few and far between. But the Pedersons' place was near the trail that led to the Turner farm where the post office was located. It was also a stopping place for neighbours making the twenty-mile trip northeast to Macoun to fetch food supplies. Horses and drivers required a respite, especially in the winter months.

~ A Celebration of a Century ~

Mary's resiliency, her ability to embrace the capacity to withstand endless chaos and repression, hastened her adjustment to living with a family of six in a sod house on the open and desolate prairie. She was an inspiration to the neighbouring women who struggled to meet family needs one day at a time. Mary was a hostess; she blossomed with the opportunities to open her humble home to her friends.

Preparing meals was difficult when there wasn't adequate food on hand. The garden area had to be broken, the soil worked thoroughly, planted and weeded. Eventually vegetables were produced. Dairy products were also important to the diet. When the homesteaders first arrived, a cream separator was not a sod house item. The milk was strained into containers and set aside until the cream rose to the top and could be skimmed off. The purchase of a cream separator at a later date brought jubilation. Cream for the coffee, even if it was homemade barley coffee, was available. Turning the separator by hand took time. Washing the separator was also a tedious daily job but performed faithfully. There was no way to keep food at a cool temperature until a well or a cistern could be dug. Lowering food in a container down into those hand-dug craters required a strong arm, a dependable rope and an airtight container.

Butter wasn't always available for the table. A quick supply could be made by shaking cream vigorously in a jar until butter formed. It was then washed to remove the buttermilk, and salt was added. Mary usually used a tall tin can and a homemade wooden dasher to churn the cream into butter. Later, she owned a barrel-shaped churn operated by a crank that turned the barrel round and round. Making butter could be a long process as the temperature of the

~ Mary Pederson ~

cream was a crucial factor. The Pederson family took turns manipulating the crank on the churn. Fresh butter and buttermilk were the rewards.

Beef cattle were butchered; meat was canned, or cut into strips and dried to make Mary's well-known *spekeKjott*. Cured meat was wrapped, hung in a dry place to cure and then sliced for the table and school lunches. Ground meat became Mary's special meatball dish. Canned meatballs were ready for unexpected company. No visitor was turned away hungry from Mary's house.

Washing clothes was an entire day's work. Carrying the water from a slough in the summer months or melting ice and snow in the sod house in the wintertime were tasks that had to be completed the day before the big washday. There were fortunate days when the rain barrel had collected sufficient water for the washing operation. Mary's copper boiler was filled with water and heated on the McClary stove. The warm water was then transferred into a tub. Mary scrubbed the clothes on a washboard, white clothes first, coloured clothes next and Charlie's overalls with all the other dark clothes last. The scrubbing board was a wooden frame with a ridged and corrugated surface of glass. This was a back-breaking task. White clothes, many made from white flour sacks, were boiled with soap in the boiler on the stove. The clothes were rinsed and carried outside to dry. Mary wrung out the wet clothes by hand. Underwear, blankets and coats were very heavy when wet and required excessive strength to wring out the moisture. Drying clothes in the wintertime resulted in manhandling frozen garments that were brought into the little sod house. Frozen large garments had a tendency to crack when folded. But Mary was convinced that freezing the clothes made them whiter and softer. Twine clotheslines in the house occupied the entire living space. However, everyone appreciated the fragrance of freshly frozen clothes that filled the home.

The women made their own soap from fat and lye. Pork fat was rendered and then strained so that it would look clear. This was usually done in the oven. It was watched carefully so it wouldn't burn. A can of lye was poured into a crock. Two and one-half pints of lukewarm rainwater were poured over the lye. Four pounds of rendered fat were placed into a second crock. The fat was slightly cooler than the lye solution. The lye was added to the warm fat and stirred. The entire mixture was poured into a large pan to solidify. It was left for a few days before cutting into bars. Mary continued to make her soap from time to time, regardless of the boasted superiority of store soap.

Ironing the clothes, with flatirons that were heavy and made of metal with wide polished surfaces, was an entire day's work. The iron had a bar to which the handle could be attached. The handle had a locking device with a wooden handgrip. The cook stove was cleaned in order to ensure irons wouldn't mark clean clothes. A steady fire was kept going to keep the irons hot. Any sheets or blankets that had been washed were folded on the table to make a smooth place on which to iron. Much of the clothing was made of cotton. Clothes had to be sprinkled the day before with water and rolled up tightly. White clothes, such as men's white shirts, would often have to be ironed on the inside as well as the outside in order to achieve a smooth surface. Ironing was a delicate operation.

Mary brought with her the skill of making Scandinavian foods. She received that knowledge as a young girl in

Norway. Her speciality was potato *lefse*, a speciality that she continued to make as long as she was capable of standing at her stove and turning the potato cakes. Her Norwegian rolling pin, neatly grooved and kept meticulously clean, faithfully rolled one *lefse* cake after another. The McClary stovetop shone as it baked each cake.

Mary's Potato *Lefse*

10 large potatoes
6 tablespoons butter
1 cup sweet cream
Use 1/2 cup white flour for every cup of mashed potatoes
1 teaspoon salt

Method: Boil potatoes, mash fine, and add cream, butter, and salt; beat until light and then cool. Add flour and roll into a ball of dough, kneading until smooth. Form into a long roll and slice into pieces about the size of an egg. Take a portion and pat it on a floured board. Sprinkle the *lefse* with flour. Roll each piece round on each side as for piecrust and as thin as possible. Add a sprinkle of flour if it sticks to the board. Use a wooden stick—a stick from a window blind, flattened at one end—to carry it to the top of the warm stove. Bake until it is light brown, turning it frequently so it doesn't scorch. When baked, place between wax paper to keep them from becoming dry. Grandma served it with butter, cinnamon and sugar with a meal or with coffee. She cut each *lefse* in half and rolled it up before serving. Birthdays, Christmas and Easter celebrations were not complete without Grandma's Potato *Lefse*.

Flat bread was a popular treat for the homesteaders. It had no moisture left in it after it was baked, so it kept well without refrigeration. It was served with meals and enjoyed as a snack at any time. Mary always had flat bread on hand.

Mary's Flat Bread

3/4 cup melted shortening
1/4 cup sugar
1 1/2 cups buttermilk
2 cups oatmeal
3 cups flour
1 teaspoon soda
1/2 teaspoon salt

Method: Pour buttermilk on the oatmeal, then mix the rest. Roll thin, place on a cookie sheet and bake in 400 degree oven until it is slightly brown, dry and completely crisp.

Let the flat bread cool. Break into serving pieces. Keep in a dry and cool place. Flat bread tastes very good spread with butter.

Everyone who knew Mary had heard about her syrup cookies. They were soft, fat and delicious. Sick neighbours, bachelors and her family waited for the baked goods that were created within the walls of the Pederson sod house.

Syrup Cookies

1 good cup sugar
1 cup shortening
1 cup syrup
Scant teaspoon salt

~ Mary Pederson ~

4 eggs
4 level teaspoons of soda—put in a bowl
1/2 cup hot coffee poured in the cup used for the syrup
Pour the coffee over the soda.
Flour to roll 1/4 inch thick—about 5 cups
2 tablespoons vanilla
6 drops anise oil

Method: Mix all the ingredients. Let the batter stand overnight. Roll out thick and cut the cookies with the lid of a quart jar. Bake for seven minutes in a hot oven—about 360 degrees.

Mary's *Favourite Recipe Book,* given to her on April 3, 1926, contained the following:

Setting a Fine Table: A silence cloth should always be used under the tablecloth to protect the surface of the table and to muffle the noise. The tablecloth should be laid smoothly and evenly on the table. If the tabletop is finely polished, doilies may be used instead of a tablecloth for breakfast, luncheon or informal dinners. Twenty-five or thirty inches is allowed from plate to plate. Lace doilies should be arranged under glasses on the service plates. Silver is laid in the order of its use from the outside toward the plate. The napkin is to the left of the silver, with the fold at the top, the open edge at the right toward the edge of the table... Remember a water glass is always essential.

Cooking Vegetables: Cook turnips 40 minutes to one hour, beets 1 to 2 hours, spinach 20 minutes, onions, boil in 2 or 3 waters until tender, spring beans, boil 2 hours, cabbage, boil 1 to 2 hours and green peas, boil 20 minutes. Always put vegetables in boiling salted water, 1 teaspoon salt to 1 quart water.

Mary faithfully fulfilled her duties as she baked, sewed and kept house. She knew, as the Sunday pillowcase message expressed, that life could appear beautiful, but she also knew by experience that her role was a life of duty to her Lord, her family and to her community. However, Mary had the ingenuity, though possessing little in material goods, to patch together quality living with only a sod house, a wood-burning stove and hard work. She was the queen of her home, her family and her community.

Charlie Pederson

Charlie was a man of courage. He faced the future with faith, purpose and stamina. His workday consisted of back-breaking duties. It was necessary to prepare the ground before he could spread the seeds of grain over his newly obtained soil. This he did with a horse-drawn walking plough and a disk plough. Thrashing in the fall was done with a scythe. Building up a farm was slow work when money and help were limited.

Life was greatly influenced by weather conditions. Winter brought freezing temperatures. And with winter came the need to adequately heat the sod house. Wood was scarce. Cow chips were used but didn't keep a fire going during the night without attention. Children were sent outside with sacks on their backs to fetch more cow chips. Desperate pioneers burned twisted hay. Lignite coal, obtained near Estevan, Saskatchewan, was preferred for heating. A lump of coal could smoulder all night. Stories have been passed down of settlers burning oats when the coal supply dwindled.

~ *A Celebration of a Century* ~

But the long distance, a sixty-mile return trip, to obtain a ton of coal took two to three days with a wagon and horses. Neighbours travelled together when possible. A pick and shovel were used to break up the coal in the small underground coal mine. The coal was loaded by hand into small carts that were pushed to the mouth of a tunnel and loaded onto the settlers' wagons. It was back-breaking work.

Pioneers found the trip to Estevan unbearably cold when temperatures were sometimes 30 degrees below. The driver would often tie up the reins, get out of the wagon, and let the horses go at their own speed while he walked to keep warm. Stories are told of blizzards overtaking Charlie. He let the horses lead the way. They never failed him; they always arrived at the door of the Pederson barn.

An abundant water supply was a necessity for a family that relied on their livestock for their livelihood. Water was difficult to find. Charlie had a well drilled, to a depth of four hundred and fifty feet, that supplied the Pedersons with an ample supply of soft water. Many neighbours obtained their water from the Pedersons' well during the thirties.

Charlie Pederson worked diligently to build up his homestead. A granary of his creation remains on the farm today, a monument to the pioneer who built it. He was a firm promoter of organizing a church and school in the community. Charlie was an active member of Lac Qui Parle Church. He took his turn acting as a trustee on the Maple View school board and he was a municipal councillor. A certificate of naturalization confirms that Charles S. Pederson became naturalized as a British subject on October 30, 1908. The *Cambria History Book* reports that he became a charter member of the newly organized Trinity Lutheran Church in Torquay, Saskatchewan, in 1915. The

~ Charlie Pederson ~

Pederson family only had five miles to travel to church each Sunday![1]

Mary carefully kept several books that belonged to Charlie. His name was stamped in each book, followed with the title of "Jeweller." The address given was Bromhead, Saskatchewan. Reports confirm that this hamlet, ten miles from the Pederson homestead, did have a jewellery store. It has been speculated that Charles had an aptitude for working with watches.

An invitation, carefully preserved by Mary, requested the receiver to attend a reception in Minneapolis, Minnesota. The invitation, neatly folded, has been kept in the original envelope. Why was Charlie invited? Was Charles Pederson recognized as a watchmaker? We can only ask.

> *The Commercial Club*
> *of Minneapolis invites you to meet*
> *Vice President Theodore Roosevelt at a*
> *Reception to be tendered him*
> *at their Club Rooms,*
> *Tuesday afternoon, September third,*
> *nineteen hundred and one,*
> *from two until five.*

The *Cambria History Book* doesn't mention Charles S. Pederson after the 1915 entry. There are many questions that remain unanswered. He left Mary and his four children and returned to live by himself in the humble house that the family had left behind in Appleton, Minnesota. Mary was left alone to manage the homestead with only her children—Arnold being thirteen years old—to help her. It has been said that the children had a strong attachment to their father. Arnold, as a wee lad, would walk in the furrows behind the plow as Charlie worked the land. He walked in the footsteps of his father. But the father left his family.

Neighbours have said that Charlie returned to the farm for a short visit. His gracious wife baked her syrup cookies, *lefsa* and flat brad, packed them well and sent the baking back with her husband when he returned to Appleton to live by himself again. The Minnesota relatives in the area sent food to Charlie's little house periodically. He became ill. Grandpa eventually returned to his wife and his homestead in southern Saskatchewan as a sick man. Mary lovingly nursed him until his death on June 27, 1927.

One of the last records available that makes reference to Charles Pederson was a Saskatchewan Pool Elevators duplicate of a $54.20 cash ticket #E501909 received from the C. S. Pederson estate, July 2, 1928. The grain was delivered to the Pool in Torquay in 1927. P.O.Vinge signed the document. The homesteader was laid to rest in the Torquay Cemetery. Mary Pederson had "At Rest" printed on his tombstone.

Notes

[1] Fifty and Over Club, *Our Prairie Heritage* (Altona, Manitoba: Friesen Printers, 1978) p.100.

Facing the Fear of Nature

The homesteaders who came to barren land and gave it life also had respect and fear for destructive natural occurrences. One cannot imagine the fearful sight of an approaching prairie fire. Thick dark smoke and thunder-sounding, cracking flames speeding across an empty prairie, overtaking anything in its path, struck terror in every homesteader. Prairie fire smoke was known to be so dense that the sun was completely darkened.

A dark cloud resembling smoke brought the prairie dwellers to action. A strip of land was dug and turned over to stop a fire from spreading. Making an adequate fireguard often required several furrows to be made, and then strips of grass were burned between the furrows to protect buildings and haystacks. However, strong winds easily influenced the size and speed of a fire. There were times in dry seasons that neighbours worked day and night to fight the prairie fires. Large holes were often left in the ground after a fire. New growth in these areas was delayed for several years.

~ A Celebration of a Century ~

The plight of Mary and other women who were often alone on their homesteads caused anxiety. Their husbands were often gone for several days with horse and wagon to fetch supplies. The women were constantly on the watch for fires. Reports speak of a severe prairie fire in southern Saskatchewan around 1906. It came with a strong north wind. Men were known to hitch themselves to plows to make furrows in the path of the oncoming fire. Sacks were gathered and water containers were filled. A sudden switch in the wind, when the fire was almost under control, sent the furious prairie fire travelling in another direction.

A full moon rising over the horizon cast an uncanny red glow on the blackened earth when a prairie fire was rampant. There was an air of unreality and apprehension.

Flames would advance with each gust of wind. They snapped and crackled in the dry grass, shot up as they devoured tall, dry brush and lowered their aim as they caught flower stalks bowed down and withered by driving winds. The Pedersons recognized and understood the eerie sight in the red western sky.

But there were men who promoted prairie fires. There were homesteaders who, in their haste to break land, tried to burn off grass on the virgin soil before breaking it. Their ribbon fireguards were often inadequate. A prairie fire easily took off and ravaged miles and miles of untamed grass, consuming unprotected homes and haystacks in its path. Hence rain and calm winds were the welcome weather reports that enhanced the peace and well-being of each homesteader.

Pioneers rejoiced when it rained, but severe lightning storms brought fear. Heavy, dark and threatening clouds that gathered quickly often brought vivid lightning flashes followed by thunderous crashes that reverberated between the clouds and earth.

Mary dreaded lightning storms. She taught her family to close doors and window lest a draft through the house attract a bolt of lightning. The vulnerability of being in a small sod house out on the lone prairie during a severe thunderstorm was daunting. Lightning flashes at night illuminated the house with an eerie glow. Fear mounted when the clap of thunder coincided with a harsh flash; the fear of a burning roof was especially great for the mother of the house when she was responsible for her family. Clashes of thunder rendered the vast sky whenever it stormed. Pioneers caught in storms a distance from their homes had nowhere to go. Trees were few and far between. Each lightning flash, like a wartime flare, would for an instant illuminate the entire heavens. It often seemed the thunder crashes would topple the thatched roof of the house onto the heads of the Pedersons beneath it. The thunder's roll, as the echo of it moved on, rumbled into the vastness of the open sky. Hearts were calm as lightning flashes grew dimmer and disappeared into the horizon. Winds subsided. There was peace, and there was sleep.

But torrential rains also brought devastation to the homesteaders' homes. Sod houses suffered under the torment of severe cloudbursts that came with force. Some pioneers used cow dung mixed with straw to give their homes a strong veneer surface after it was exposed to the sun for a long period. Such a finish was known to suffer if strong wind-driven rain was excessive. The odour and colour of such damage wasn't appreciated.

It wasn't easy for Mary to keep the contents of her home dry during a devastating rainstorm. There was nowhere to turn for relief. A heavy downpour of rain could begin, and in moments the floor would be covered with streams of water. The rain would beat in, driven from every direction. Bedclothes quickly became wet when water dripped through the thatched, vulnerable roof.

However, after a storm had subsided, opening the door brought nature's reward to compensate for any devastation and discomfort. A tranquil sky was a beautiful sight. A full double rainbow and the sun piercing through the clouds, casting its broken beams over a landscape that earlier was dark and threatening, were a joy for the pioneer to behold. Treetops glistened with gold. Seeing a rainbow arch in its splendour across the sky left a comforting impression on the people who had just experienced the anxieties of a violent storm. The rainbow was a gift from the Creator, a gift of encouragement, promise and affirmation.

But there were times, perhaps only twice a year, when the pioneers looked for an awesome and entertaining performance in the sky. Such were the times when the northern lights danced across the heavens. Northern lights could be seen in southern Saskatchewan in summer or autumn. Grain thrashing would often go on after dark in the fall of the year. The northern lights were especially enjoyed at that time, a reward for hard-working pioneers.

The Worn Black Purse

Mary's old clutch purse contained her treasures. Church-related papers, including tracts and meditations, were in the purse. The church was important to the Pedersons, so they welcomed the organization of Trinity Lutheran Congregation in Torquay. The town was only five miles from their farm. The family became charter members of the new congregation. Arnold, then fourteen years old, was in the first confirmation class. Mary was instrumental in helping to organize a Ladies' Aid group and became their first president. She had previously borrowed nine dollars from her husband to buy cloth. Mary sewed articles that were sold at an auction sale. The money raised was used to purchase a lectern and benches for the church basement. The church building was completed at a later date.[1]

The purse's aged contents also revealed that Mary was a friend to many. She was not only loved by her family but was appreciated dearly by all who knew her. One letter tucked in the lining of her purse confirmed the high esteem with which she was held in the community. The letter, written in Norwegian, was from a neighbour who had also moved from Appleton, Minnesota. The receiver must have treasured this letter, kept for years. Mary received the letter shortly after Charlie left his family and returned to Minnesota:

Torquay, January 18, 1916

Dear Friend, Mrs. Pederson:

Be so kind as to excuse the way I am. I ask you to excuse me, as I'm not able to be along in the Ladies' Aid. I really thought it was my duty to belong but I understand what I can manage.

I don't manage to do any work outside of the home but I must say to you thank you very much for all the cozy times. It has been cozy for me to sit in on the meetings and I think that you do so much and I marvel over all your kindness and energy. It has been a great joy for me to have seen your good smiles and heard your friendly words.

I was moved and pained in my heart when I heard that your are sad. We can't expect you to always smile. But God's children are often disappointed and maybe it was good for something.

We hear people don't always smile for joy. It is a big relief to see happy people to go with their happy work. Many thanks for all your Christmas gifts that you gave to all and to many.

Many thanks for all the cozy times.
That's all for now.

Mrs. L. Bergum

There were additional papers, yellow with age, tucked

in the worn black purse. A neatly folded piece of parchment paper was slipped under the clear window of a small brown case that held "Mary Pederson's National Registration Certificate." The form read:

To All Whom It May Concern

This is to certify that from evidence submitted before me, I am satisfied that Mary Pederson of Torquay in the Province of Saskatchewan, wife of Charles S. Pederson is a person naturalized as a British subject by operation of law, who, but for such naturalization of for any disability contained in the Naturalization Act, 1920, is qualified and would be entitled at the date of this certificate to personally naturalized in Canada.

Dated at Estevan, Sask. this 6th day of October, 1920
Signed by E. B. Mylie, Judge of the District Judicial, District of Estevan.

Further investigation of the frayed case revealed another piece of parchment.

It was a receipt made out to Mrs. C. Pederson for the sum of two dollars for a "Family Plot, No. 9, in Trefoldighed Cemetery" in Torquay. Grandma Mary always planned for tomorrow. She had carried this receipt in her purse for many years.

Another folded paper, also yellow with age, was small and insignificant. The tattered edges led a person to believe that it was scrap of paper of no value. Further investigation revealed that a Norwegian message, though ink had faded, was still readable. It was a prayer; perhaps the prayer she prayed at her Ladies' Aid meetings:

~ *A Celebration of a Century* ~

Oh God, Thank you for your great care
and for all you provide for us.
Bless now these gifts so that we can
receive strength from them.
Thank you for giving us heavenly bread.
Give peace and the fruits of your spirit to our land.
Protect our Christian heritage. Bless the work we do.
We ask that you give us good health.
Bless us all with the living bread that Jesus
bought with his dear blood.
Amen.

Notes

[1] Fifty and Over Club, *Our Prairie Heritage* (Altona, Manitoba: Friesen Printers, 1978) p. 475.

The Red Velvet Album

Precious items carefully preserved in Mary's glass-doored cabinet were intriguing to young grandchildren. The fat velvet photograph album with the tarnished gold clasp captivated the greatest attention. This album, estimated to be over 125 years old, had been repaired with white thread. Frayed edges of the padded covers revealed the many times eager eyes had toured the pages of the picture history of Mary's family.

A family of fifteen children can have numerous special events. Confirmation and wedding pictures portrayed a few of the important dates in the lives of the siblings. There were pictures of Sigrid, Lars, Sisssel, Ivar, Ingeborg, Anne, Knut, Andres, Young Lars and (Birgit) Bertha, and their families. Photos of Kari's family left behind in Norway were also in the album. Kari was the second "Kari" of the family. She was named after an older sister who had died in infancy. Portraits of Mary's confirmation, another of Mary clad in a Norwegian costume from the Hallingdalh area and a wedding photo of

~ *A Celebration of a Century* ~

Mary and Charlie are included in the picture-gallery book.

Johannes was another one of Mary's brothers. She proudly displayed pictures of his visits to the Pederson farm. John, as he was later known, was born in 1870. He immigrated alone to America in 1886. A two-storey house made of large cement bricks, individually patterned and moulded, still stands proudly today in Minnewaukan, North Dakota. The bricks have never cracked. The inside of the house boasts intricate woodwork, carved mouldings and a graceful stairwell. The house was built for Bertha, a girl he couldn't forget back in Norway.

John had longed for a home and family, a home that would be his own, which was impossible for him to obtain in his homeland. A young lady in Hol had seen the eighteen-year-old lad, suitcase in hand, walking alone to the station that would eventually lead him to the open sea. John later contacted the girl, who had never forgotten him. But Bertha had a duty to care for her ailing mother. In time, a new bride and an ailing mother-in-law immigrated to America and moved into the home that John built!

In the front of the album are pictures of another special sister. Margit, born in 1883, was nine years younger than Mary. She left Minnesota and moved to Torquay. She decided to help Mary, knowing that taking care of the homestead plus caring for her children was a heavy load. Margit was an enthusiastic, vibrant sister who brought sunshine to the Pederson home. She found a job as a housekeeper for a single man in the community. She completed her housework duties in the day and helped sister Mary with her chores each evening. Margit volunteered her time in church life. People who remember her speak of a talented young lady who inspired the young and old alike.

~ *The Red Velvet Album* ~

Margit was highly praised in the community. She was featured in the *Cambria History Book*. It states that the Rural Telephone Company built a central office in Torquay in 1916 The book continues to say:

> A Central Office was erected, "A building 14 by 20 feet with 10 foot poles was to be erected with specifications as follows: Outside covered with lap siding. Inside to be lathed and plastered, double floor and flat roof, Tenders were let for Central Operator. Miss Margie Knutson's tender was accepted at 49 dollars per month. Central hours were to be from 8 to 1 in the forenoon and 3 to 6 in the afternoon with evening hours of 7 to 9 on weekdays. Sunday the office was to open from 8 to 9 in the morning and 6 to 7 in the evening. The operator was to sleep in the office and answer night calls for the doctor….Long distance was to be connected so the central hours were changed to open 7 in the morning to 10 at night on weekdays and on Sundays and holidays to be open from 10 to 12 in the forenoon and from 4 to 6 in the afternoon….Miss Knutson was to be allowed to purchase Kelsomine for the office and to do the work herself. A linoleum was to be purchased for the floor."[1]

The cheerful visits from sister Margit didn't continue for long at the Pederson homestead. The *Cambria History Book* records that the directors' meeting on December 1, 1920, was one of sadness as they were advised of the death of their central operator, Miss Margie (Margit) Knutson. The secretary reported that he had sent a wreath of flowers for her funeral and also sent the following message on behalf of the company:

~ *A Celebration of a Century* ~

In honour of our true and faithful employee, Miss Margie Knutson, do we as a company tender this as a last token: Resolved that in the death of Miss Knutson we have sustained the loss of a friend, whose fellowship it was an honour and pleasure to enjoy, and our heartfelt condolences go to the remaining relatives, over whom sorrow has hung her sable mantle.[2]

Mary Pederson again experienced loss. She accompanied her sister's body by train back to Appleton, Minnesota. Margit died of quinsy, an acute infection located between the tonsils and the pharyngeal constrictor muscle. There were no antibiotics, no treatment available for a dear aunt and sister.

There is a clearly marked grave in the north end of the Torquay cemetery. The single man who had first given Margit employment was laid to rest several years after her death. His grave stands alone. There is no evidence of additional family members buried nearby. Was an early death hastened due to a broken heart? Were dreams unfulfilled? There is only speculation. But everyone knew that Margit was a blessing to every life she touched.

The picture at the back of the album evokes a keen sense of sadness in the viewer. A young woman and her four children are standing near a white casket. There are many friends and relatives surrounding the family. Sigrid, Mary's eldest sister, born in 1859 in Hallingdal, Norway, married Ole Berg in 1877. The youngest of her four children was Ole Olson, who lived near the Pederson place. (Refer to chapters 1 and 16.) Ole Berg passed away in 1884 at twenty-nine years of age. Sigrid and her family immigrated to America with her father four years later. As stated in chapter 1, Sigrid married Halvor Robertson two years after she arrived in Minnesota. Sigrid later gave birth to four more children.

~ The Red Velvet Album ~

Sigrid took a trip to Norway; then World War I broke out. Sigrid was not allowed to leave Norway for a year and a half. Son Knut Olson, born in Norway in 1881, lived on the farm in Minnesota but was a photographer in Dawson. Knut had previously had a photograph gallery in Canada, and was in the process of building a gallery of his own in Dawson. He passed away on July 12, 1915 while his mother was in Norway. He was thirty-four years of age. He had gone up to the hayloft in the evening to get feed for his animals and was stricken with severe pain. His sisters thought Knut had left for town. The next morning, his groans attracted the attention of his sister. Knut was conscious and in severe pain. He passed away later that day. A letter was written conveying the sad news to his mother in Norway. There wasn't need to send a cable to Sigrid; she couldn't leave Norway. A letter carrying the sad message was sent across the ocean.

Hence the red album became alive as people featured within were given names, voice and purpose. It brought back memories of Mary speaking of her family with pride and longing. It connected the viewer to the noble line of strong people that stretched back for generations but, at the same time, still walked beside the beholder. Their stories must live on, their stories of triumphs, challenges and heartaches. Each person is a link in the chain. The red velvet album has told us so....

Notes ──────

[1] Fifty and Over Club, *Our Prairie Heritage* (Altona, Manitoba: Friesen Printers, 1978) p. 24.

[2] Fifty and Over Club, p.25.

~ *A Celebration of a Century* ~

*Erwin Henry Pederson
in his youth*

Erwin Henry Pederson

Eldest son Erwin was a fun-loving member of the Pederson family. He was "big brother," and with that prestigious position came great expectations and responsibilities.

He is remembered as the lad who made everyone laugh at school. He was known to disappear under a desk when the teacher left the room. But Erwin possessed a gift that gave him recognition—a gift that enabled him to fix any item that was in need of repair. His hands were skilled in whatever he chose to do. A trombone has remained on the Pederson homestead for years. It sits in an intricate trombone case complete with lining, clasps and gold decor. Erwin built the case over eighty-five years ago.

Erwin was only eighteen years old when he and a friend struck out with a team and a buggy to look for a homestead in 1913. *Happy Valley Happenings*, the Big Beaver, Saskatchewan, history book, states that the young men drove to what is now the Big Beaver district. Erwin filed on a homestead. He built a two-room shack with lumber hauled from Viceroy, Saskatchewan. He spent the required number of months to prove up a homestead. He periodically worked at a coal mine in the area. Ambitious Erwin had obtained a steam engineer's licence back in 1914. He later purchased an eight plough furrow steam engine, went back to Torquay each spring, broke up land with his steam engine and then thrashed grain with his own thrashing machine until freeze-up. He ploughed up half of his community, perched on the platform of his sharp steam engine.

It was 1916. Erwin had found that it was lonely living in a small house all by himself. He ventured away from his home one day on a bicycle, determined to bike across the border to Whitefish, Montana. His bike had a flat tire on the way. He figured he could get help at a farm a short distance from the mishap because he saw a bicycle leaning against the house. This was the home of Lucy Wager, a young maiden, who later became his wife. They were married in Weyburn, Saskatchewan, on June 12, 1918. Erwin's homestead home was no longer lonely. Lucy and Erwin were blessed with five children: Glen, Vern, Lucille, Kenneth and Donna. The history book also states that with each addition to the family came another addition to the house. Erwin could fix anything needed.[1]

In later years, the Erwin Pederson family moved from the farm to the town of Big Beaver, a town that boasted five elevators, set in a beautiful valley. Erwin continued farming and worked at custom repairing of machinery, work that he enjoyed.

The eldest Pederson brother passed away on November 3, 1974. He was laid to rest in a graveyard that is set in rolling hills in a peaceful and picturesque area south of Big

Erwin Henry Pederson

Beaver. The quiet cemetery stands as a monument to courageous and hard working people, like Erwin, who loved the community they built.

Notes

[1] Big Beaver and District, *Happy Valley Happenings* (Regina Saskatchewan: W.A. Printworks, 1983) pp. 325-326.

~ A Celebration of a Century ~

1899
Ida Josephine Pederson

Ida Josephine Pederson Hagen

Ida Pederson was born on March 2, 1898, in Appleton, Minnesota. It was said that Ida was "born with a silver spoon in her mouth." She was a petite, pretty girl with many gifts. Music was her life. She could carry a tune as soon as she learned to talk. Grandma Mary marvelled at her ability to harmonize with others as a child. She had a true alto voice. Ida and brother Arnold started to sing duets at an early age.

Ida, as a teenager, was one of the first organists at Lac Qui Parle Church and later at Trinity Lutheran Church. The *Cambria History Book* also tells us that she was one of the charter members—as was Mary—of Trinity's Lutheran Ladies' Aid. Ida assisted in the organizing of the LDR, Lutheran Daughters of the Reformation, or *Pega Frening*. It was organized sometime between 1918–1920 at Trinity in Torquay. The first president was Aunt Margit Knutson. This ambitious group purchased the first church bell. They bought material for the pulpit, baptismal font, seats and altar that were built by T. O. Kvammen and are still in use today. The LDR bought the altar painting, a picture of Christ knocking at the door, for seventy-five dollars.[1,2]

Ida enjoyed giving music lessons. She led a Bible study and taught girls how to do handwork. The sale of their articles made many purchases for the church possible. Ida also taught Sunday school, and parochial school in the summer. She loved people.

In 1926, Ida married Chris Hagen. A daughter, Phyllis, was born December 12, 1927. Ida and Phyllis moved back to live with Grandma Mary several years later.

Ida became seriously ill when she was thirty-six years old. She was rushed to Ambrose Hospital in Ambrose, North Dakota. Ida's appendix had ruptured. She passed away on the operating room table during surgery. The date was June 11,1934. The deserted hospital still stands today. Circular cement steps, overgrown with moss and fern, are steps Ida struggled to climb years ago. She was never to walk back out that hospital and down those steps again.

Phyllis was five years old when Mary had to become her mother as well as her grandmother. Mary, in her sixties, introduced Phyllis to school, sewed her clothes, cared for her through childhood illnesses and did her best to be the mother Ida would have been had she lived. Mary faithfully displayed her love for Ida's daughter Phyllis as she took Ida's place as a mother in Phyllis's life.

Chris Hagen, Phyllis's father, passed away in 1967 at Hudson Bay, Saskatchewan.

~ Ida Josephine Pederson Hagen ~

A Picture from Ida's Life

The year was 1927. Aunt Ida and Uncle Chris lived in their farm home south of Torquay. Ida was a natural hostess. She had the ability to make any area cozy and comfortable. She enjoyed people, and people loved her.

Ida enthusiastically prepared for a great occasion when relatives arrived from the United States.

Aunt Ida is the third lady standing to the left, with four daisies on her jacket. She could create any type of handwork with her gifted hands. She crocheted the flowers and planted them on her garment. Mary's sister, Annie, is standing to the right of Ida. She was born in 1872. (She married Charles Lantz. They had three boys.) Mary is next, and her sister-in-law, John's wife Birgit, is standing on the far right. Chris Hagen, Ida's husband, is the man sitting on the left in the front row. Mary's brother Johannes (John) is the third man to the left. (Refer to Chapter 7.) Arnold is sitting next to him. He doesn't appear to be paying attention to the photographer; it looks as if he is having a conversation with his brother, Clarence. Kermit, John and Birgit Knutson's son, is sitting on the far right. He was born in 1915. (He married Eileen Peterson, and they had three boys.)

It is probable that the car parked in front of the home belonged to Ida and Chris Hagen. These prairie people, dressed in their finest, took time to enjoy one another.

Notes

[1] Fifty and Over Club, *Our Prairie Heritage* (Altona, Manitoba: Friesen Printers, 1978) p. 475.

[2] Over Fifty and Over Club, pp. 102-103.

Clarence Selmer Pederson

Clarence Pederson was born December 26, 1899, in Appleton, Minnesota. A neighbour girl who frequently cared for Clarence remarked that he was the prettiest baby she had ever seen. That babysitter eventually became Clarence's mother-in-law. He was four years old when his family moved to Canada; he learned early to herd cows and gather cow chips. He was eight years old before the first school classes were organized at a neighbour's home. Clarence was a hardy lad who adapted quickly to the life of a homesteader.

The second youngest son of the Pederson family had the ability to captivate the attention of the people in his life, especially the hearts of young people, The advent of the car age fascinated Clarence. He became acutely knowledgeable of vehicles with four wheels, which were replacing the horse and buggy. Young people enjoyed the opportunities to ride with Clarence in his black Essex, the car he loved and frequently overhauled. He was popular with the young and the old, loved by all who knew him.

Clarence was first employed in the twenties at the Eika Grocery Store in Torquay.

He then ran an implement business in Cornach, Saskatchewan. He later returned to Torquay, where he again worked in the same store, now called the J. B. Johnson Store. Nieces and nephews claimed Uncle Clarence's affection. They received special gifts each Christmas from a special uncle. And Clarence's store intrigued the youngsters, the store with shelves lined with popcorn, peanuts and chocolate bars.

There wasn't much money for buying food during the thirties. Farmers exchanged their produce for groceries. They came when egg crates were full; cream was churned and formed into wrapped pounds of butter. Nieces and nephews were fascinated by the store-experience stories Uncle Clarence related to them. One such story concerned a distraught farm wife who rushed to Clarence with two pounds of her butter. She told him she had found a mouse in her churn when she was making butter, She asked to exchange her butter for two other pounds of butter. The clerk immediately went to the back room, removed the original wrappers and put new wrappers on the lady's butter. She happily left the store. Anything could happen at the J. B. Johnson Grocery Store in Torquay.

Clarence was married to Myrtle Gysler in 1939. They moved to Fillmore. In 1943, the couple moved to Beaubier, Saskatchewan, where Clarence operated his own grocery store, managed the Co-op and was postmaster. They now had a family of three—Donna Lou, Lyle and Doris. Clarence sold his store in 1948, due to ill health. The family moved to Estevan, Saskatchewan, where Clarence worked

Clarence Selmer Pederson

for Tisdale dry cleaners. In 1954, they moved to Regina, where Clarence worked for Rainbow Cleaners.[1] Clarence passed away in Regina on April 21, 1966.

Notes

[1] Fifty and Over Club, *Our Prairie Heritage* (Altona, Manitoba: Friesen Printers, 1978) p. p. 476.

~ A Celebration of a Century ~

1926
The Young People at the Pederson Farm
At the top of the hayloader are Horace Johnson and Clarence Pederson.
Bottom row, L to R: Gudrun Vinge, Clara Vinge, Agnes Vinge,
Arnold Pederson, Beatrice Johnson, Edna Bergum and Albert Vinge.

Arnold Glen Pederson

A. G. Pederson was born February 7, 1902, in Appleton, Minnesota. He was a two-year-old when he came with his family to establish a homestead on the endless prairies. There had to be little comfort or shelter for Mary and her toddler when she had to plant a garden, bake bread, can meat, sew clothes, do laundry, iron, make butter, clean the sod house, plus other daily obligations. Conditions necessitated an entire day to complete each task. But Mary was a courageous, sturdy woman who worked diligently with strong and willing hands. Her young son had a great role model.

Arnold, being the youngest, saw many changes during his childhood. Maple View School was organized and built by 1908. Arnold was six and old enough to walk just over a mile to school. The sod house was replaced with a two-storey house that boasted shingled walls and roof. They no longer had to travel fourteen miles to Lac Qui Parle Church, as a new congregation was formed in Torquay, five

~ *A Celebration of a Century* ~

miles from the homestead, in 1915. Arnold was in the first confirmation class.

Arnold, thirteen, and Clarence, fifteen years of age, were left to help their Mother and Ida farm the land after Grandpa moved away from the family in 1915. Erwin had already departed to start a homestead of his own. Cows, pigs, turkeys and chickens required constant care. They provided food for the table. Horses also required care; they provided labour for farming.

Arnold became seriously ill when he was seventeen years old. He developed pneumonia. The doctor was afraid that he might not recover. Despite December temperatures, the doctor insisted that his bed be kept outside in the cold. The doctor knew of no other way to treat the acute illness. The patient lived and was never ill again until the last years prior to his death in 1981. Arnold had to make one trip to the dentist in his lifetime, at age fifty-four. He promptly fainted after having one tooth pulled.

The *Cambria History Book* speaks of Arnold as being ambitious as a young lad. He acquired an agency to sell bicycles. He only sold three of them, but he was able to get a discount on his own. He was intrigued with the advent of earphones used to listen to a radio. He sent for radio parts and assembled them. Arnold purchased a bellow-type camera, made his own darkroom and developed his own pictures. He took a penmanship correspondence course and later, with Albert Vinge, attended a business college in Minneapolis. He worked in a hardware store for his Uncle Knut Knutson in Appleton, Minnesota, in 1925–1926. Arnold worked with Knut's son, Austin, who later moved to Minneapolis and became vice president of First National Bank. The stately Knutson Hardware store still stands in

~ Arnold Glen Pederson ~

Appleton today as a monument to its industrious builders, Mary Pederson's relatives.

Arnold met Beatrice Johnson through church functions. Beatrice lived fifteen miles south of the Pederson farm. Courtship dates included trips to a circus and to a silent movie in Estevan. Beatrice's young sister and two nieces had to go with them. Arnold drove Beatrice to Ambrose to buy an ice cream cone on many occasions. He patiently waited for Beatrice to agree to marriage. Beatrice had made the comment to her mother that she wished Arnold wasn't already showing signs of going bald. Little sister Viola told Arnold she would have him if Beatrice wouldn't make up her mind. But Beatrice did care. She preserved a card, tinged yellow with age, in her photograph album. It was dated March 18, 1926. Arnold's neat and precise printing on the card stated:

> *Therefore, when we build let us think that we build forever. Let it not be for present delight, nor for present use alone. Let it be such work as our descendants will thank us for and let us think, as we lay stone on stone, that a time is to come when those stones will be held sacred because our hands have touched them, and that men will say as they look upon the labour and the wrought substance of them, "See! This our fathers did for us."*
>
> *Fond wishes for a Happy Birthday to the Sweetest Girl in the world.*
>
> *From Arnold.*

There are courtship pictures that tell a story. Pictures in the antique family album portray various young-people gatherings that took place at the Pederson farm. A hay loader rack, standing by the granary, is featured on one picture. Beatrice and Arnold, in their courting days, are sitting on the rack surrounded by their friends. Pictures of the couple taken in Minneapolis are also in the album. Arnold had persuaded Beatrice to leave her telephone operator job in Torquay and attend Bible school in Minneapolis with him in the winter of 1927. They both found this experience to be one of the highlights of their lives.

Beatrice and Arnold were married in Beatrice's home church, Salem Lutheran, on October 29, 1927. It rained so heavily that the bridal couple couldn't leave the bride's home after the reception. The honeymoon trip to Plentywood, Montana, had to wait until the next day. Their record of registration of marriage, licence # 87451, has yellowed with age. The writing is faded and blurred. Specific questions included on it ask if the groom and bride could read and write. The registrar, A. G. Vinge, signed the document.

Arnold, with Beatrice at his side, took over the Pederson farm. Charlie had passed away in June of the same year. Mary, Ida and Clarence purchased a house in Torquay. Arnold had saved $1,000 to spend on replenishing the brown shingled house. It paid for a hoosier kitchen cupboard, oak china cabinet, oak table, three matching chairs, bedroom set, library table and rocking chair. Beatrice was pleased to display her hope chest articles on her fine furniture. She made a house a home.

~ *A Celebration of a Century* ~

October 29, 1927
Arnold Pederson and Beatrice Johnson's Wedding Day

Mrs. Arnold G. Pederson

The late twenties brought many changes on the Pederson homestead. It wasn't always easy being a farmer's wife in the late twenties. This was southern Saskatchewan where the grass blew, blew as a restless sea tossed by waving winds. The prairie setting stretching on into the horizon gave hope to a young heart, hope of happiness as vast as her vision of the distant landscape. But there were lonely hours for Beatrice in this time of adjustment that protruded themselves like the sprinkling of small stones that made their sharpness known underfoot.

Days moved into weeks and weeks melted into months. Beads of perspiration flowed freely on washdays spent scrubbing on the scrub board. Ironing days had to be the day the bread was baked; flat irons had to be kept warm. When the coal supply was low, gathering cow chips took time, but the black cookstove consumed them with a fury. Days could be long. Mounds of ashes cast off by the McClary stove gave evidence of the busy housewife's labours.

~ A Celebration of a Century ~

Anticipation grew as Beatrice and Arnold awaited the birth of their first child. The ground floor of the shingled house didn't have a bedroom. It was December. The upstairs bedrooms could be cold when they were only heated by two thinly-cased stove pipes, one from the heater and the other from the kitchen stove, that protruded through holes in the ceiling. The expectant father used a rope to lower their double bed through an upstairs window to the frozen ground below. This bed claimed the area on the first floor that bathed in the warmth of the iron-clad coal heater. The rocking chair had to find a home in the kitchen.

The time for the birth arrived. Arnold hitched Lady and King to the sleigh and brought Dr. Smythe out to the farm. The Pedersons' humble living room witnessed the mother's pain and the baby's first breath.

The first newborn had arrived on the Pederson homestead. But Grandmother Johnson's reply to her daughter Beatrice's expressions of joy for her new baby was simply, "Even the mother crow thinks her baby is the greatest."

Beatrice remembered, too, seeing Dr. Smythe later at a ball game. She wanted to show her baby to her doctor, but listened to an older sister. "Do not bother him, he sees so many babies."

The upstairs metal bed made two more trips through the upstairs window and down to the ground below. Clinton and Verna were also born in the family living room in the cold of approaching winters. Arnold wasn't at home when his wife knew Verna was soon to be born. She sent an urgent call on the only existing telephone, a "wire fence telephone," that enabled a conversation with the Bergums. The contact was made but the doctor didn't arrive. Arthur

~ Mrs. Arnold G. Pederson ~

Bergum had fallen off his horse and had broken his leg in his rush to fetch Dr. Smythe. The baby didn't arrive until the doctor came. Beatrice was in bed ten days after each birth. Her babies thrived well, fed as her mother ordered:

"Cook oatmeal for hours until a fine gruel gathers at the top and feed it to your baby."

A letter Beatrice wrote to Mrs. Horace Johnson, her sister-in-law and her bridesmaid, who lived fifteen miles south of the Pederson farm, reads:

December 15, 1929

Dear Clara:

Sylvia received such a nice birthday card from you, Thank you very much for it. I am going to put it in her baby book.

I have been wishing that I could have come to visit with you for awhile. I was on the way the day of the ladies' aid, but we turned again on account of the roads being so heavy. I had planned on staying until Monday and staying awhile at each place, but it did not work out to be that way. Hope it can be later.

Sylvia is fine. She is trying to say a few words now, but she does not walk yet. She can walk along the couch when she holds herself onto it. She gets good use out of those mittens you gave her. She even wears them at night. We sleep upstairs, and she does not like to have her arms in under the covers.

I do hope that you are feeling well. You are a good friend. I do so much wish that you can be happy.

I have been kind of busy, been sewing some, and I quilted on a big quilt and one for Sylvia. Arnold made a little bed for Sylvia. It got nice. We do not get up before eight in the morning so the days seem to be so short. We were at church today and afterwards we went to see Grandma Pederson.

I would like so much that Horace and you would come up for Christmas if it would be possible then. I wish you folks would try—if you don't think the ride is too long.

I have not much to write about, but I thought I would thank you for Sylvia's card and I also thought you might like to hear from me. May God be with you and Horace. It is nice to know that the Lord is nigh unto all that call upon him.

Your Sister,
Beatrice.

A 1928 Picture Story:

Pictures have a way of bringing memories alive. The picture of Mary, with neighbour Hilda Salte and Beatrice, was taken in front of Mary's house in Torquay. Arnold took the picture the year after he and Beatrice were married. Ida, Mary and Clarence had only lived in town for a short while. They had not yet painted the trim on their house. Arnold's Model T car, his pride and joy, can be seen parked in front of the house.

The stately elevators of Torquay stood straight and tall in the background of the picture. They gave Torquay class and marked the town with distinction. They were a tribute to the pioneers who built them and used them. Beatrice's brother-in-law, Peder, was agent for the Saskatchewan Wheat Pool elevators. He handled the grain that came from the Pederson farm.

~ Mrs. Arnold G. Pederson ~

Dr. Smythe's house is also on the picture, to the left of the elevators. Dr. Smythe was the doctor in Torquay for many years.

From left to right: Mary Pederson, Hilda Salte and Arnold's wife, Beatrice

The Turbulent Thirties

The depression hit southern Saskatchewan. Dust storms and drought struck in the early 1930s. Soil on the Pederson farm drifted like snow after a snowstorm. Dirt rolled in windswept waves across fields that had produced bountiful crops. The fields were riddled with gopher holes. Water was scarce on many farms. Neighbours depended on the Pederson's deep well for their water supply.

The Pedersons hoped for rain each time a dark cloud formed at the horizon. Their hopes vanished when a dark, dirty cloud enveloped the farm in darkness. Sometimes the storm would last for an hour, and other times the winds would blow and cause the soil to drift for days. Farmers would often reseed parts of their fields. The drifting soil would cut off tender young plants. The field would often look barren after a severe dust storm. Fine dust would gather along fences banked with Russian thistles, the only weed that thrived in the drought.

Arnold, like his neighbours, struggled because of the drought, insect invasions, wind destruction and ultimately the depression. Various aspects of farm living were affected, including transportation. The curtain-flapping 1925 Chev that had been purchased from Arne Vinge had been replaced with a 1927 Model T before the drought. But cars take gas, and there no longer was money for fuel. The Ford was put in the granary until the family could afford to drive it. Arnold's two oldest children took the opportunity to fill the fuel tank with wheat. The distraught father attempted to flush out the grain but to no avail. There wasn't money to repair the fuel tank. The engine was taken out and used to operate a feed grinder. Arnold ground oats for the neighbours and for his own needs. The Model T was never on the road again.

Arnold was dependent on his horses for farming. Russian thistles didn't provide an adequate diet. He was forced to ship Florey, Lady, Trixie and Queen to Manitoba for the winter of 1930. Horses staying in the area became too frail, skeletons protruding, to put in the spring crop. There were neighbours who left their homesteads behind, and there were men who left their wives at home to care for the farm while they went north to find temporary work. The government helped the needy by shipping out vegetables, dried fish and canned goods by rail.

Arnold and Beatrice struggled through those years. They were dependent on their cream, butter and eggs to buy groceries; but eggs sold for five cents a dozen, four gallons of cream sold for forty-five cents and a pound of butter was worth ten cents. Their ingenuity and determination saw them through the depression years. They

~ *The Turbulent Thirties* ~

always believed that next year would be better. Their land had produced bountifully for them before, and they knew it would do so again. They never forgot how to pray.

~ *A Celebration of a Century* ~

Maple View School Students—1912
Back row, L. to R.: Clarence Pederson, Luella Grubs,
Edna Bergum, Gladys Smith, Agnes Brownhill
Middle row: Lucille Youngberg, Robert Hall, Nellie Hall,
Arnold Pederson, Arthur Smith, Leonard Bergum
Seated: Clara Bergum, Ruth Magoon, Anna Smith, Alice Youngberg

Maple View School Days

Life went on during the depression. Protective parents ensured that the Pederson children never suffered. There was church to attend, which meant a five-mile trip by horse and wagon. There were family gatherings where creamy potato salad and scrumptious meatballs were served. There was the Norwegian Society that met in the farm homes on Sunday afternoons. Trying on ladies' hats in the upstairs bedroom while their owners listened to the Norwegian service below was only one of the amusements for the younger people on afternoon outings. There was always time for a birthday or anniversary gathering. People had time for each other.

Somehow Maple View School teachers always stayed at the Pederson home. The west upstairs bedroom was called the teacher's room. Beatrice had to cope each day with the pressure of preparing adequate meals, eating on time, keeping children in check, feeding the chickens, gathering eggs and milking cows when Arnold worked late in the fields. She usually was scrubbing clothes on the washboard until after school had finished on wash days. She received fifteen dollars a month for boarding the teacher. One cheque was spent to purchase new blankets for the teacher's bed.

But there was a teacher who wished to live in the teacherage at Maple View School. It was a small partitioned-off area in the northwest corner of the schoolroom. It was furnished with a woodstove that didn't have a warming closet, three apple boxes stacked together for a kitchen cupboard, a table, two chairs and a cot with wings that made into a double bed. It was decided that the eldest, now of school age, should stay with the teacher in the school. The teacher didn't want to live alone. The eldest had been looking forward to the day when she could walk to school, lunch pail in hand. She would have to wait....

The six-year-old was intrigued with the enormity of the school, the multitude of desks, large walnut libraries and walls that were many feet long. But nightfall came quickly, and with it came looming shadows and pangs of loneliness. The floors creaked and footsteps echoed as she made her way through the darkness of the school to use the outdoor facilities. She never wanted the teacher to know that her pillow was damp from tears each morning. But the grade one pupil soon realized that noon hour was a highlight each day. Individually marked potatoes were put in the ashes of the furnace at recess. It didn't matter that the outside of the potato was charred and black. The potato's interior was great with butter.

Memories of early school days frame pictures in the mind of the fascination of learning, but they are peppered with pangs of loneliness experienced each day after school. The

~ Maple View School Days ~

six-year-old watched her friends, swinging their lunch pails, leaving for their homes. Her job was to break sufficient twigs off the trees to heat soup for supper. Tears flowed as she tried her best to break off stubborn and stiff twigs that scratched and poked. But there was always Friday, the day she could walk home from school. Waiting for Fridays became easier and her pillowcase drier as the joy of learning deepened.

In later years, being able to walk to school brought the joy of walking home with the teacher to find Beatrice's lunch of newly baked cookies or cream and bread topped with chokecherry syrup. Living with teachers became a way of life. The chair at the south end of the kitchen table belonged to the teacher. Arnold's place was at the other end. Beatrice sat near the cupboard while the six-year-old sat next to the teacher. She didn't always appreciate the responsibilities that came with being the eldest of the family.

The family has vivid memories of the young man that periodically drove into the Pederson farmyard in his car— with flapping curtains on its windows—to see a teacher. Little heads would peer through the upstairs windows in an attempt to get a glimpse of the teacher and her beau, regardless of Beatrice's warning to stay away from the window. They found themselves overwhelmed with awe and wonder as they were introduced to the fascination of romance.

Family members know now that the teachers who touched their lives left a great imprint. They inspired, encouraged, and challenged them. The teachers cared. They taught their students to be diligent in all their endeavours. Arnold and Beatrice saw the importance, despite sacrifices, to board a teacher in order to ensure their children's education. Invasion of family privacy was but a minor selfish undercurrent in childhood memories of living with teachers. The example of the teacher, in the home and in the classroom, introduced lessons that wove naturally and comfortably into the fabric of everyday life.

There were cold, blizzard days when Arnold drove the teacher and his children to school with the wagon and horses. Neighbours consistently helped each other. The Salte caboose picked up a passenger from the Pederson farm in the thirties. The trips went well when the caboose remained upright. The Pederson farm became a refuge for stranded neighbours in need. One evening, the Saltes were late in leaving a social at Maple View. It was dark, but their horses knew the way home, despite the narrow road. A severe lightning storm ensued. Rain came down in torrents. The family steered their wagon down the lane to the Pederson place, or the "Comfort Inn" as the Saltes called it, where they stayed for the night. Neighbours needed each other.

The school Mary and Charlie help organize was a hum of activity on special occasions. Social life thrived. People drove for miles to attend the Christmas concerts at the school. Students practised for hours. Everyone knew his or her part in the program by memory. Fathers would put up a platform stage, and sheets were hung on a wire and used as curtains. Pie socials and box socials also attracted a large community attendance. Picnics held at the school drew young and old alike. Maple View School was the social hub of the community.

The school was closed in 1953. The barn and school building were moved away. There is no indication today that the school was located at North East 33-2-11.

The Stately Barn on the Pederson Place

Charlie's barn was a landmark, a prominent building that did its best to give the Pederson farm a sense of class. This barn was regarded as a special legacy, but it was revealing its age. Dark red paint applied to it always had a miraculous way of covering up aging scars, especially when viewed from a distance.

To young children, the barn was a place worthy of respect and honour. It was the home of Queen, Florey, Trixie, King and Lady—Arnold's horses that faithfully tilled the soil on the farm. Chickens periodically laid their eggs in the mangers. Newborn calves claimed a fenced-off corner. Arnold trained the cats to drink fresh-from-the-cow milk in direct-stream fashion. Children waited for this procedure as much as did the cats.

The ladder to the upstairs loft opened up a new world of wonderment to a young child. Mounds of hay were castles and valleys. Storybooks came to life in such a setting.

Dreams could almost take wings in the barn loft haven.

Children were easily entertained in such a milieu. Playhouses took shape and schoolrooms were built in minutes. An accidental happening brought everyone quickly to reality when brother fell through a floor opening. A manger filled with hay gave him a welcome landing below.

At harvest thrashing time, the barn had a major role. Teams of horses filled the barn each evening; their owners slept in the hayloft above. The barn was bursting at the seams, but there was also activity in Beatrice's kitchen. Hungry thrashers consumed copious amounts of roast chicken, vegetables, bread and pie. Lunches with hot coffee were regularly carted out to the field. Weary men and horses returned to the barn each evening.

The barn was used as a teaching tool. The Pederson children soon realized that misbehaving had consequences. Arnold led them to the barn door after one such incident. He showed them the holes left by nails that he had removed from the door years back. The nails were gone. The children were forgiven, but the act left scars. They caught on....

There are many memories from growing-up years that centre around barn activities. Nesting hens, sitting on their eggs in apple-box nests, were kept in the hayloft until their baby chicks hatched. Any clucking hen found about the yard was carried up the loft ladder and given eggs, with three weeks to prove she would do her faithful-mother duties. Discovering new kittens in the mounds of hay also brought joy to any beholder. New life on the farm enhanced daily living for the young and old.

The Pederson cows were an important part of the farm family. Daisy and Brownie both thought they owned the barn, but they had distinct personalities. It was impossible

~ The Stately Barn on the Pederson Place ~

to milk Daisy if you had pins in your hair. Her swinging tail caught every pin. Brownie never took her eyes off anyone who was milking her. The cows demanded to be treated with respect.

Family Gatherings

It was a big event when Mary's sisters and brothers came across the United States border to visit at the Pederson farm. Father Arnold harnessed Berdie and Queen to the mower and cut the grass around the shingled brown house. Beatrice canned a good supply of meatballs, made jars of rhubarb jam and preserved mustard bean pickles.

Mary was busy baking in her house in Torquay, too. She made giant-sized syrup cookies, doughnuts, potato *lefsa* and raisin bread. She sewed herself a new print dress and apron. Mary didn't see her siblings often. She was delighted to welcome them to her home in town and to the farm.

The Knutsons were a merry group who laughed, chattered, teased and reminisced. Arnold put extra boards on the kitchen table so everyone could sit down and enjoy his wife's meal. Beatrice made the tastiest meals on the black McClary stove, meals fit for a king. Everyone knew that these visitors were just like royalty, because that is what Mary called them.

~ *A Celebration of a Century* ~

There was great excitement at the farm the day Arnold took a picture of everyone. Mary's sister Belle Ellefoson stands on the left, then Mary, her sister Bertha Dahl, Beatrice holding little sister Verna, Clara Olson, Arnold's cousin's wife who lived a mile from the farm, Mary's brothers Knut and John, Arnold's cousin Ole Olson and Knut Knutson's wife, Carrie. Clinton and Sylvia are standing in the front row.

Mary also welcomed the relatives to her home in Torquay. They feasted on her Norwegian baking at coffee time and on her mashed potatoes and casseroles at mealtime. Mary was frugal in every way, but she still could prepare delicious food for her table. She knew from experience how to stretch anything of value. Her neighbour once remarked that she probably sharpened her pencils over the fire in the cook stove. Grandma could take a joke any time.

Entertaining wasn't always easy for Mary. It was difficult to keep food from spoiling. She had a rope fastened to a wooden orange box that she lowered into her well. Her

Family Gatherings

milk, cream, butter and meat had to fit into the small container. But she was never without food. No one left Mary's home hungry.

Mary's brother John is standing on the far left; her sister Belle Ellefson stands next to Mary, then her sister Bertha Dahl, Ole and Clara Olson (her nephew and his wife), her son Clarence and her brother Knut Knutson.

Mary's sister Belle Ellefson, born in 1870, was married to Ellef Ellefson, who also came from Hallingdal, Norway. Ellef and Belle had ten children. Sister Bertha (Birgit) was born in 1883, the youngest of fifteen children. She married Arnold G. Dahl. They had fourteen children. Knut Knutson was born in 1875. He married Carrie Jacobson. They had four children.

Ole, The Homestead Neighbor and Cousin

Ole and Clara Olson, featured on the previous family-gathering pictures, moved next to the Pederson farm in 1913. Ole was a son of Sigrid, Mary's oldest sister. He was born in Norway in 1884 and immigrated to Minnesota with his widowed mother in 1888. (Refer to Chapter 1.)

Ole married Clara Bjorklund. They lived in Ambrose, North Dakota, three miles from the Canadian border and fifteen miles south of the Pederson place, for a few years. A son, Marvin, was born in 1912 in Ambrose. He passed away when he was one year old. Ole and Clara moved to Canada shortly after their son's death. They built a house and barn on land neighbouring the Pederson homestead. Mary and her family encouraged and supported them in every way they could.

Clara and Ole spent much of their time at the Pederson farm. Clara had to walk around a large slough to get to the Pedersons, but she never complained. Often a horse and buggy could be seen out the south window of the shingled brown Pederson house. The buggy was the means of transportation when both Ole and Clara were game for an outing. Coffee time at the Pedersons' was the reward.

The Pederson children remember Ole's dog named Flubb. Ole taught the dog to growl each time someone spoke to him in Norwegian. Norwegian-speaking visitors were not always told why the dog growled in their presence.

Christmas was special at Ole and Clara's home. Clara made a delicious meal for her neighbouring family each Christmas. Norwegian baking was in abundance. Everyone went home with candy in his or her pockets.

The Olsons moved to Estevan, Saskatchewan, in the forties. Clara passed away a few years later. Ole remarried after Clara's death and moved to Melfort, Saskatchewan.

He became ill after his second wife's death. Ole Olson's file included a receipt for a $6.00 ambulance ride to the hospital, and also several telegrams that Arnold sent to Ole's family. The first telegram, at a cost of $1.40, sent to Ole's sister Emily Hanson in Dawson, Minnesota, early in the morning on June 27, 1961, reads:

Mrs. A, M. A. Hanson
Dawson Minn.

Ole passed away Monday at midnight. Undecided about funeral. He has plot at Melfort but Torquay or Estevan more convenient for relatives. Would like you to decide. Wire me at Torquay if coming and when.

Arnold Pederson

Ole's sister sent a reply at 9 a.m. the same morning to her cousin Arnold:

Ole, The Homestead Neighbor and Cousin

Arnold Pederson
Fone 14r14
Torquay Sask.

Letter on way. Unable to come. Do as convenient there.

Emily Hanson

Arnold had notified the family of Ole's illness. Portions of letters from his sisters—Emily Hanson from Dawson, Minnesota, written the day Ole passed away, and Ingeborg Mikkelson, following Ole's death—stated:

June 26, 1961

Dear Folks,

Your letter and telegram arrived about two hours apart. Sorry to hear about Ole, although, being at the age he is, it didn't shock us too much, as he has complained a lot about his health lately. Then, too, there is no doubt in my mind but that he is a child of God. I understand that he is very low and that it is a matter of time and we are all so helpless back here. His full sister Ingeborg "Belle" Mikkelson, who lives in Minneapolis, is soon eighty-two years old and not in shape to travel. Brother Robert is laid up with a bad case of bronchitis, and I know full well I couldn't be able to travel that distance, as my health is not exactly good. I feel so badly; it's awful to be so far apart in a case like this when we are all the age we are and in delicate health. I am seventy-two years old now and commencing to feel it....

Your Cousin, Emily Hanson

~ A Celebration of a Century ~

July 10, 1961

Cousin Arnold,

I want to thank you first thing for what you have done for us. You sure did well to take care of everything. God bless you for all you did for Ole... It is too bad when one is so far away. I couldn't think to come out there. I've been doctoring this spring. I am still taking medicine. But I can do my housework so that is pretty good anyway for eighty-one years. Martin is eighty-seven. He is making picnic tables and cement sidewalks so he sure is a strong man....I miss hearing from your Mother since she passed away. We were such good friends. God will reward you for what you have done for us....

Mrs. M. E. Mikkelson

P. S. My husband made and sold two picnic tables last week.

~ *A Celebration of a Century* ~

A Trip to the Neighbours
The Arnold Pedersons' mode of transportation in 1946.
Blanche, Beatrice, Arnold and Clinton are in the wagon.
(Beatrice is holding baby Lorna.) Verna and Sylvia are on the sleighs.
The shingled brown house Charlie Pederson built in 1912 stands
in the background. The 32-volt wind charger that was put up
that year is perched on the roof.

Introducing the Forties

The 1940s brought various changes to the Pederson place. Many neighbours had given up and moved away to greener pastures during the drought of 1930–1939. People who had suffered through that era and remained were rewarded by the return of bountiful crops in the early forties.

The Pederson family had welcomed a new member of the family in 1938. Blanche was born in Clara Ruland's home in Torquay. She lived across the street from Dr. Smythe. Her house became a convenient maternity home. The double bed upstairs in the Pederson home stayed where it belonged. The farm living room wasn't required to be a birthing room again.

A letter typed on a yellow page of parchment bears a distinguished letterhead. Mary wasn't finished handling estate affairs:

~ *A Celebration of a Century* ~

SASKATCHEWAN POOL ELEVATORS LIMITED

Regina, Sask.
18th March 1940

To: Mrs. Mary Pederson
Torquay, Sask.

Dear Madam: Re Estate File No. 107

We understand you are the owner of certain Elevator and Commercial Reserve Wheat Pool Deductions, formerly belonging to Mr. Chas. S. Pederson who is now dead.

Arising from surplus earnings over a number of years our organization has a certain sum of money standing to the credit of certain of its patrons. With this sum it has decided to invest in the purchase of Elevator and Commercial Reserve Deductions belonging to deceased Pool members. We hereby offer to buy at the price of 50 cents on the dollar of the face value of both deductions belonging to you as heir. Offer must be delivered to our office within 45 days of this date, after which our offer will not be open for acceptance.

Saskatchewan Pool Elevators Limited
Geo. W. Robertson, Secretary.

The year was 1941. The faithful horses received respite. A 1928 John Deere Model D tractor on steel wheels was purchased. The tractor was showered with attention. The marvel of a mechanical workhorse filled the farmer's heart with awe and wonder. The tractor served well until 1947 when a new John Deere Model D was purchased. Arnold also purchased a second-hand John Deere Model AR with four wheels, a row crop model, at a later date. The two front wheels were close together. Reports of this tractor tipping on occasion caused the family concern. But it stayed upright and proved to be an efficient lightweight tractor, especially when it came to hauling grain.

With productive crops came the ability to finance a car. In 1942, Arnold purchased a 1928 Chev, a black car with window blinds on the back window and running boards that enabled easy entry into the car. The car was a luxury to a family that had depended on horses and wagon for transportation. The car served faithfully for several years. However, the steering mechanism malfunctioned on a trip to Estevan. Arnold suddenly lost control of the car. It dived into the ditch, up the embankment and into a dry slough before eventually coming to a stop. There were prayers of thankfulness expressed. No one was hurt.

The aged binder had served well but a one-year-old Allis Chalmers combine was purchased in 1942. The binder was kept as a portion of the crop was thrashed each year to ensure a supply of straw. A Cockshot #7 Co-op combine was purchased in 1947. Stooking was an art taught on the Pederson farm at an early age. No one appreciated the combine more than a brother and a sister. They had kept their sandwiches and cookies in the first stook made, marked distinctly with three grain bundles stacked precariously on top of it. More time was spent trekking to the treat-pinnacle than in erecting stooks.

There are distinct memories of unusual weather conditions in the forties. March of 1943 is remembered as a month of severe blizzards and excessive snowfall. There wasn't train service from Estevan because the railway was blocked in various areas. When the train did start out from Estevan, a snowplow and an extra engine went ahead of the train.

A Celebration of a Century

There was a crew of nineteen men that travelled to Torquay that day. They shovelled out the railway tracks when it was required. Farmers had to take extreme caution during severe blizzards. Arnold tied binder twine to the house door, unwound the twine as he walked to the barn to do chores and secured the end of the twine to the barn door. He found his way back to the house by following the twine on several occasions. Winter could be harsh, but the farmers embraced the capacity to move forward despite extreme cold.

Arnold's files include various data concerning farming operations. A portion of a letter from a landowner who operated Headington Motors, Inc., in New York read:

549 W. 49 St., New York
September 21, 1943

Dear Mr. Pederson:

I am taking steps to have the title on the Canadian land that you rent transferred to my brother and myself who are the sole heirs... I am enclosing $87.00 which is in accordance with your invoice rendered and covers the breaking of 15 acres of ground @ $4.00 per acre as well as 9 bushels of seed flax furnished by you... What are your thoughts on breaking additional land this next year and planting flax or wheat?

Yours very truly,
E. W. Headington

A second letter in reference to the rented land concerns Arnold's request to purchase the land:

~ Introducing the Forties ~

Sanbornville, New Hampshire
November 28, 1944

A. G. Pederson:

Dear Sir: I have your letter dated Nov. 24, 1944 in relation to the land of which I am part owner situated near your farm. I am only one-fifth owner of the land. It came into our possession through our having been endorsers on a note given back by a former owner who failed to pay the bank in full. Personally I am interested in your offer and I have today seen three of the other owners face to face. They have seen your letter. The fifth owner, Lawyer Wm. N. Rogers, is out of town. I will contact you after I see him.

Very truly,
J. H. Garvin

The above land was purchased, seeding was undertaken and the garden was planted in the spring of 1945. June brought a newborn to the Pederson family. Lorna was born in St. Joseph's Hospital in Estevan. Three years went by and Linda, also born at St. Joseph's, joined the family in June of 1948. There was always room for extra chairs at the round table even if a teacher consistently sat at the far south end.

Inside activity after sundown on the Pederson farm was totally dependent on a clean kerosene lamp chimney and a functioning gas lamp. It was an exciting day when Arnold purchased a light plant in 1946. A 32-volt Leland Generator was purchased for $85.70; panels, bolts and pulleys were $15.30 and the 32-volt wind charger and batteries cost $352.00. Arnold's receipts also revealed that the additional cost of freight, hauling, wire, insulators, yard light and entrance materials brought the total cost of the light plant to $427.00. The family revelled on a windy evening; lights were bright, *Lux Radio Theatre* and *Amos and Andy* came in loud and clear and the kerosene lamp was spared the nightly trip up the stairwell at bedtime. A new world opened for the Pedersons with a flick of a switch.

Community involvement was a priority for the Pedersons. Arnold served on committees, such as church boards, school boards and municipal boards. Arnold became a Saskatchewan Wheat Pool delegate of District One in 1947. He became a director of the district in 1952, a position he retained for eighteen years. He was again a delegate for two more years. He retired from this position in 1972. He spent twenty-five years with the Saskatchewan Wheat Pool. Arnold was appreciative of the men that he worked with, the new friendships that were made, and moreover, the goals and aspirations of the Wheat Pool that he upheld.

~ A Celebration of a Century ~

1940
Arnold Pederson and Carl Knutson cutting the ice from the dugout and preparing to haul it to the icehouse. Ice didn't last long in the kitchen icebox during the summer months. The Pederson's windmill and barn tower in the background.

Ice for the Ice House

Keeping milk and cream from souring was a persistent problem. Storing the cream can in the cellar was not an effective solution, but the only solution available. There was always the rush to get the cream can to town. People just accepted the situation.

Arnold purchased an old-fashioned icebox in 1940. It had a pipe leading from the drain pan, through the wall, to outside. He was determined that he could keep ice in the icebox consistently if he could create an efficient ice house. He built a building behind the house that served his purpose. The building didn't have a floor. That was necessary so water could always seep away. He first dug a few feet down and removed the soil from the area. He covered the dirt floor with sand and placed sawdust above it. The small building was moved above the excavation. His children were fascinated with the structure.

The Pederson farm had boasted a Reg Jones dugout since 1939. Arnold solicited the help of neighbour Carl Knutson, and they proceeded to cut ice for the ice house on a winter day when it wasn't too cold, so the ice could be handled more easily. An ice saw, two pairs of ice tongs and a crowbar had been purchased. The ice was cut into blocks that could have been twenty inches square. Cutting the first square was the most difficult. Arnold used a crowbar to chop it out. He measured out other ice squares. The opening in the ice made it possible to cut the other blocks easily with the ice saw. The ice blocks floated when they became free. The two men grabbed the ice blocks with their tongs and pulled them out of the water. Toeholds had been chopped in the ice close to the water to ensure the men wouldn't slip into the water. The ice squares were pushed to a plank and slid onto the stone boat. Children had to watch from a distance. It was serious work.

Packing the ice into the ice house was a tedious job. Arnold left a space around each block of ice. He filled these areas with chipped ice and sawdust. A second layer of ice blocks was carefully put into place. The blocks were insulated with straw. When a piece of ice was put into the icebox, the empty area in the ice house was carefully covered.[1]

Meat and dairy products finally kept well on the Pederson farm. The icebox served well. This piece of furniture eventually lost its natural wood finish and was painted a dull green, a colour that didn't glorify its worthiness.

Neighbour Carl continued to help Arnold with the heavy job of cutting and hauling ice blocks each year until 1948, when a refrigerator was purchased. The ice saw and tongs hung in the machine shed for years as a reminder of a dangerous, yet skilled, operation.

Notes

[1] "If You Are Selling Sweet Cream," *Country Guide*, October, 1937, p. 20.

~ *Ice for the Ice House* ~

The Arnold Pederson Family in 1954
L. to R.: Blanche, Verna, Clinton and Sylvia
Seated: Arnold, Linda, Beatrice and Lorna.

The Aladdin Windsor

Days on the Pederson place were numbered for homesteader Charlie's shingled brown house. It had withstood the dust storms of the depression, winter blizzards and summer's torrential rainstorms for over forty years. The upstairs was still cold in winter, a dirt cellar wasn't an efficient vegetable keeper, and a family of eight didn't fit into it anymore.

Plans for an Aladdin Windsor house met with keen approval. The Pedersons liked the Aladdin Windsor that stood on a farm near a neighbouring town. Arnold borrowed their plans. The Pedersons decided to build a 24' by 34' house, a two-storey spacious bungalow. It had a square-shaped veranda and a vestibule that led from it into an attractive living room. An archway to the left opened into the dining room. There was a good-sized kitchen, pantry, bedroom and sewing room. The upstairs had three bedrooms and an office in the square hallway. The new home couldn't be erected fast enough; anticipation for a house replacement ran high.

~ A Celebration of a Century ~

A blue, faded *Handy Farm Account Book* was found in a storeroom box. An accounting course that Arnold took in 1925 was closely followed, as every cent spent for the building of the Windsor must have been recorded in the account book. There were eighty-eight receipts tucked in the handbook; forty-one were from 1946, thirty-seven from 1947 and ten from 1948. The receipts added up to a sum of $2,011.39.

Ben Johnson, Arnold's father-in-law and a skilled carpenter, was the contractor.

The frame went up quickly after the basement was dug and the foundation completed. The new house grew threateningly in front of the shingled brown house that shrunk in its shadow. A cement mixer and various trucks travelled in and out of the Pederson place as different phases of the construction were completed. Entries under *Labour and Trucking* in the accounting book list a total of $750.00 given to several individuals, including Ben Johnson, for their services. Entries included $5.00 paid to the truck driver who brought out the cement mixer, $7.00 to the nephew who brought out the bricks and $5.00 to another truck driver for hauling cement.

As the house took shape, rooms bloomed into distinctive personalities all their own after the bare partitions were properly clothed. The Aladdin Windsor's dignified double-mullioned windows gave grace to the dining room and living room. Doors and closets, cupboards and flooring and several coats of paint brought finishing touches. The gables, brown-shingled trim and stucco finish gave the exterior an attractive appearance. The Aladdin Windsor was completed. The shingled brown house was moved to Torquay. It was sold for $600.00; the contract stated that Arnold would be paid $25.00 a month. The Windsor now reigned as queen of the Pederson place.

~ The Aladdin Windsor ~

The Pederson Farm in 1966
The Aladdin Windsor has replaced the shingled brown house but the barn, granaries and chicken coop still stand. The ice-house, to the left of the Windsor, no longer robbed the dugout of its ice.
It had become a brooder house.

A Visit from Brother Andres

Mary Pederson's brother Andres and his wife visited Torquay on several occasions. He was eight years old when he immigrated to America with his mother, Mary, and six of the youngest brothers and sisters. He was an ambitious young man who built his own furniture store in a small town in North Dakota. He was concerned because there wasn't a funeral home in his community. He purchased a book and studied it. He became a self-taught funeral director. He built a funeral home as a lean-to on his furniture store.

Andres had many interesting experiences. He worked closely with the local doctor. They had a good working relationship. An Indian reservation in the area gave Andres an urgent call on one occasion. Andes picked up the doctor, as was the usual routine, and they drove out to the Indian reservation. A friend was sitting near the body when they arrived. The doctor pronounced the man deceased. The friend insisted that he accompany the body back to the

~ A Celebration of a Century ~

Knutson Funeral Home. He refused to leave the premises once they transferred the body into the morgue. The friend spent the night in the waiting room. Andres returned to the funeral home the next morning to find the devoted friend feeding the supposedly deceased man toast and coffee. Life was always interesting for Mary's brother Andres.

Andres and his wife, Clara, out visiting the Pederson place. Beatrice is standing on the far left with Mary's brother Andres, holding a doll, beside her. Blanche is next and is standing with Clara Knutson, who holds another doll. Mary stands in the front with Linda on the left and Lorna on the right. Arnold took the picture. He consistently enjoyed taking pictures.

~ *A Celebration of a Century* ~

Wesley Engen sitting on his Grandfather Arnold's combine in 1960. The faithful windmill towers above it. Wesley passed away November 12, 1991.

The Fifties and Sixties

Time marched on. The Pederson farmyard took on a new decor. A yard fence was put up around the house, Three hundred spruce trees, two to three inches high, were planted. These trees were supplied free for the asking from the Saskatchewan government. The farmer faithfully worked the soil around the trees periodically. He manoeuvred the hand plow as one of his teenagers sat on horse Berdie and steered down the long lanes. It was hard work. Arnold and the rider's trousers, and the horse, were damp with perspiration after the weekly workout. And the trees grew.

The interior of the new house was equipped with furnishings that had been ordered from Eaton's catalogue. A receipt from the Canadian Pacific Railway confirmed that they charged $8.74 collect for shipping a chesterfield, an armchair, four cushions, a kitchen table and chairs and four thirteen-foot congo linoleum rugs. An additional receipt states that a sink, at the cost of $7.50, was also ordered from

Eaton's catalogue. The Windsor was home. Life went on.

Mary Pederson had keenly followed the building of the new house that replaced the shingled brown prairie homestead house that had at one time been hers. She now lived alone in her home in Torquay. The house was old. It could be drafty in the cold of winter. She retired late every evening during the colder months. A large chunk of coal had to be put in the heater at a late hour to ensure a fire would be burning in the coal heater in the morning. Mary was getting older. The responsibilities of keeping up a home and grounds weighed heavily on a woman who had been independent and efficient all her life. It was time to make new living plans.

A letter in Arnold Pederson's files came from Rev. K. Bergsagel, Administrator of the Lutheran Sunset Home in Saskatoon, Saskatchewan:

October 29, 1956

Dear Mr. Pederson,

We shall be glad to keep the room for your mother until she comes on November 3rd. She may feel free to take along the articles you suggested. If she has a little radio she may also bring that if she wants to—or any other little article which may help to make her room more homelike to her.

She will find congenial company here and it is my hope that she will feel at home.

Sincerely yours,
K. Bergsagel

Mary did find congenial company in the Sunset Home. Her letters, written in Norwegian, confirmed her happy acceptance of her new home and new friends. She arrived

~ The Fifties and Sixties ~

in late fall, so Christmas at the Sunset Home was a new experience for her:

December 28, 1956

Dear Beatrice and All,

I must try and write. First of all, hearty thanks for everything. Many thanks for the big box of Christmas gifts and goodies you sent.

First I must tell about the Christmas party yesterday at 3 in the afternoon. We were fifteen Norwegian ladies. We were in the sun parlour. We sat. They asked, "Sister Mary" to say grace so I prayed after we had eaten. We sang Christmas carols.

You must know that you got a lot of praise for your baking. People liked it and it all went. They brought coffee from the kitchen when I served your baking. I wanted to pay for it but they said it was a treat. I didn't have to pay. All thought it was so cozy.

I said one time that they should have prayer meeting once a week. Some of the ladies have started it in the wings and then others could do the same. But they said that they meet in the evenings when the students meet for prayer meeting.

All of December was good with many Ladies' Aids bringing food and candy. Sunday schools came with programs, too.

We had a special Christmas Party in the evening; it was cozy to have real fun. Supper at 5 and after that there is a short service. It comes on the loud speaker so we listen to that (those who can hear). Afterwards, they put gifts under the tree. There were three parcels for each one. I didn't think I'd get anything because I'd come so recently but I got a handkerchief, finger mitts, writing paper and a box of candy. I got sixty Christmas cards. I've never had so many cards before.

There was a service here on Christmas day. Jacobson had the service—good speaker. Storaasli had a record of the Christmas service at the seminary and he came and played it for us.

I must write about myself. I do not have any pain in my chest now. That pain I had in my thigh made it hard to sit in a chair. I walk better now but I have to use a cane. I gained three pounds. I eat well and I thrive.

I can't be lonesome here. Clarence called me on the phone. Many thanks to Verna for the nice brooch. I used it at Christmas. Thank A.G. Vinges for the gift they sent me from T Eaton Co. and to the Ladies' Aid in Torquay. They sent me a candy box.

It was God's Spirit that led you to send me all those gifts and goodies. You must not think I did much work for the party when I served your baking. I only bossed the others. They were so glad to do everything.

Mother

Beatrice especially appreciated a letter that was sent from the Sunset Home in 1957:

March 18, 1957

Dear Daughter-in-law, Beatrice,

Greetings and have luck today on your happy birthday. May the Lord bless you and give you good health through your lifetime. Many thanks for your goodness.

You are like a daughter to me. I couldn't love you more than I do. I haven't anything against you. You have to excuse me but I want to let you know how I feel and thank you for everything.

We must pray for the people in our family, pray that they will all take the Christian life in more earnestness and be serious about their Christian life.

Arnold came to visit me. I was happy to see him. I had a letter from Lucy. I was so glad to hear from her. But I wish I could see better. I see so poorly.

May God give you many years to be with your dear ones.

Mother

Further letters revealed that Mary continued to rejoice in being a member of the Sunset Home family. She prayed for good health; she didn't want to miss any activities that took place at the home. However, there were letters that did inform the family that Mary Pederson wasn't consistently well:

May 10, 1957

Dear Mr. Pederson;

Enclosed herewith please find receipt for recent check in the amount of $35.00, board payment for the month of May.

You will also be pleased to know that your mother has taken her meals in the dining room today and in general feels much improved. I was able to move her roommate so that your mother now has the whole room to herself, which in itself will speed her recovery. It is also an advantage that she is happy and content and has much in common with the other ladies in her wing.

Sincerely yours,
K. Bergsage

A Celebration of a Century

Mary Pederson's last Norwegian letters to her family continued to reveal her appreciation for all her blessings despite her disabilities:

May 6, 1858

To My Dear Ones,

All is well here. We have it good. I can go to the dining room and to the bathroom with just my cane again. So all the trouble they have with me is to bring me my pills in the morning.

This will be a funny letter. It may be a mixed-up letter. It is so tough to write. I see so poorly. But I want to say thank you to the girls, Linda and Lorna, for sending me letters. It is so much fun to receive them. I had a letter from Mrs. Mork and Mrs. Knutson. I should write to them.

Edith Vinge came to see me on Thursday. She is so busy but it was so good to see her. I am always glad to see someone from Torquay.

I mustn't forget to tell you that Glen Johnson came here on Tuesday in the evening. He is studying to be a minister. He sang and he had his girlfriend along. He got me acquainted with her. They are going to get married in two years.

Now I have made my letter longer than I thought so now I have to break it off with a heartfelt greeting to all of you.

Mother

The following is the last letter Mary Pederson wrote before her death:

~ *The Fifties and Sixties* ~

May 22, 1958:

Dear Ones,

They sang happy birthday to me last evening. They came to see me. I received a large flower basket from Phyllis. Say thank you to Mrs. P. Vinge for what she sent me.

I have been in bed most of the last week. My heart is weak but I feel better now. Thank you to all of you for everything. May God richly bless you all.

Mother

Concern of the administrator and residents at the home for Mary's state of health was expressed in several letters:

July 26, 1958

Dear Mr. Pederson,

As I visit your mother from day to day I notice that she is gradually growing weaker and it may well be that this will be her last sickness before her Lord calls her to Himself. Speaking to her last night I found her very tired, but mentally alert. She asked about her friends at the Home and said that possibly she would not return to the Home. If that should be the Lord's will she would be content and look forward to the better home.

She mentioned once that now she wished she could have been a little closer to home so she could have seen her loved ones more often. But even that would have to be as the Lord wills it.

Yours in the Master's service,
K. Bergsagel.

The following letter was written after Mary's death:

~ A Celebration of a Century ~

September 5, 1958

Dear Mr. Pederson.

With our grateful acknowledgement of your memorial gift to the Sunset Home I want to add my personal thanks, first for the gift, secondly, and especially, for the privilege which was ours to have Mrs. Pederson with us in the evening of her life. Though frail in health, she was happy in the Lord and she scattered sunshine all around her. She felt at home here from the first day she came, and she never ceased to be thankful to God and her church for a Christian home for the aged.

While she is greatly missed by all here we are also thankful to God that He has taken her unto Himself and left us with the memory of a pious soul devoted to Him and to His Church.

Sincerely yours,
K. Bergsagel

September 10, 1958

Dear Mr. Pederson,

Mr. Sam Kvinlog who lives here mentioned to me yesterday that he had lent Mrs. Pederson $5.00 before she was hospitalized, and since he went back to the farm she had no chance to repay him. He said it was not important to mention it, but I assured him you wanted to know it so that he could be repaid.

Mrs. Pederson has been greatly missed by all her friends here and I deeply regret that I could not come to Torquay August 5, 1958, for her funeral.

Sincerely yours,
K. Bergsagel

The Fifties and Sixties

A list of memorial gifts, given in Mary's memory, and their donors has been kept in Arnold's file. The total amount of gifts was $115.00. "A Bill of Costs from the Surrogate Court of the Judicial Centre of Estevan in the matter of the Estate of Mary Pederson," signed by lawyer R.D. Newsome, listed the total fees and disbursements to be $134.53. Mary Pederson had gone to her eternal reward.

A letter from the Lutheran Hour in Edmonton stated:

December 31, 1958

Dear Mr. Pederson;

We can think of no better way to memorialize our departed loved ones than by sending a contribution toward the spreading of the gospel of Christ, something which is definitely done by means of the Lutheran Hour. We are sure that your now sainted mother also believed in the Christ whom the Lutheran Hour proclaims and your generous offering will help direct others to that heavenly home above.

Wishing you abundant blessings of Christ for the New Year's, we are,

Very Sincerely Yours,
W. C. Eifert

The following letter, written by brother Erwin, stated:

Big Beaver, Sask.

Dear Brother Arnold,

Received your letter a few days ago. I am just digging out of another snowstorm. We had so much snow again now as we did the previous storm. I will be satisfied if this is our last of winter snow.

We have been well all winter but snowed in most of the

~ A Celebration of a Century ~

time. We were up to Regina 15th to the 19th and I had my semi-annual checkup and all is well. We visited the family and enjoyed our time in Regina.

In regard to you plan of settling Mother's estate, I think it is wise to dispose of it as you have outlined in your letter. Whatever you do is agreeable with me.

Well, this is Monday morning, May the 1st, and I am going to try again to clear the driveway from snowdrifts.

Hope to see you folks this summer. Trusting this finds you and yours in good health with lots of ambition which is needed during a late spring.

With love,
Erwin

~ *The Fifties and Sixties* ~

April 19, 1965
Clinton Pederson and Muriel Moen's Wedding Day

The Close of a Century

Life went on at the Pederson farm. Hope and aspirations for fulfilling days grew with each sunset and each new day. The cattle that lived on the farm were decreasing in number because Arnold spent ten days of every month at Saskatchewan Wheat Pool meetings. But Beatrice wanted life on the farm. Chickens claimed the run of the Pederson place.

Arnold continued to embrace the work of the Saskatchewan Pool. Notes of a presentation he gave at a pool function read:

Future Policy

It has been said that anything that grows, whether it be a child or a tree, has a future. As long as there is reasonable growth, there is health. We do not have to dwell on the rapidity of the growth.

I've heard it said quite a number of years ago that there was a British firm that manufactured a silver polish that had been on the market for a generation or more. Its name was a standard byword. Those responsible thought it no longer necessary to advertise their product. They said as it were, "Boys, we've reached the summit, from now on we only coast." And coast they did, right into bankruptcy. The moral is simple and evident here. We just can't afford to rest on our laurels. We need to work together to achieve a common purpose. Co-operation is a way of life. Where there is growth there is life. And when there are surpluses, we must strive to divide those surpluses and there will be growth.

When we look to the future, we also do well to give a look to the past. It's well to know something about the road we came on and how far we have gone.

But Arnold had made a decision in 1970. He resigned as director of District One of the Saskatchewan Wheat Pool. A copy of the ad he put into the Weyburn Review reads:

We wish to express our sincere thank to everyone who attended the Wheat Pool Committee banquet in Torquay on October 20th. The handsome gift of a comfortable reclining chair is mush appreciated and will be a constant reminder of your generosity. The kind words and handshakes will be long remembered as expressions of warm friendship. In the twenty-five years I have been privileged to serve as Wheat Pool Delegate and Director, I have gained many friends whom I have leaned to regard most highly. You have a standing invitation to drop in any time for coffee and a visit.

Beatrice and Arnold Pederson

A letter from the Saskatchewan Wheat Pool, Office of the Senior Manager, reads:

~ The Close of a Century ~

November 4, 1970

Mr. A.G. Pederson,
Torquay, Saskatchewan.

Dear Arnold:

I have just noted from the minutes of your district meeting October 15 and 16 your announced intention to retire as Director at the close of the present term. While I was aware that you had this question under consideration for some time, it comes as a matter of regret that you will not in future be at your accustomed place at the Board table. With the numerous changes in the Board in recent years, I have felt your background knowledge and experience to be an important stabilizing influence in the decision-making process. I have greatly appreciated our personal relationship over the years, and it has been an honour for me to work with a man of your integrity, sincerity and good faith.

I understand that you will stand for election again as Delegate, so we will see you from time to time in the future. However, I would take this opportunity of wishing you and Mrs. Pederson the best of health and happiness in the years ahead.

Yours sincerely,
K. Mumford,
General Manager.

Various changes had taken place on the Pederson place. But keeping up with a change of machinery and car was only considered when there was money in the bank to pay for it. Each purchase was given considerable thought before a transaction took place. However an exception to the usual routine was made back in 1954. The family was up at 4:30 in the morning and on their way to Moose Jaw. Blanche had to catch a train that would eventually take her to the Lutheran Collegiate Bible School in Outlook. The car failed at nine o'clock in the morning near Weyburn, Saskatchewan, due to a defective oil pan. Time was limited. The big decision to buy a new car was made and the family left Weyburn in a new Plymouth Savoy shortly after ten that morning. The train waited for a half-hour for the family in the 1955 car to arrive. Blanche got to LCBI in time for her first day of school. But the ensuing years saw the family resort back to the making of more thorough and painstaking decisions at the Pederson place.

The stately barn was losing its class. It lost its purpose in life when animals no longer lived under its roof. Tearing down the barn was a difficult and long procedure. During the demolition, a rabid skunk attempted to attack Arnold. He successfully killed the animal. The RCMP was notified. A diagnostic report confirmed the skunk was indeed rabid. The barn was gone, but not even a skunk infestation could erase the many memories of the crowning pinnacle on the Pederson farm.

Arnold retired as delegate of the Saskatchewan Wheat Pool in 1972. But it wasn't until the mid-seventies that health problems surfaced. He suffered a cerebral accident in 1977. Leaving the farm was a difficult decision for Beatrice and Arnold. Beatrice had spent close to fifty years on the farm. Arnold had lived on the Pederson place for seventy years.

Arnold and Beatrice moved to Torquay, where they celebrated their fiftieth wedding anniversary. Arnold spent his last days in the Estevan Nursing Home. Arnold's days as a committee member never left him. When the residents of the home sat down at their tables and waited for meal, Arnold was known to stand up and sincerely call out, "We will now call the meeting

~ *The Close of a Century* ~

to order." There were residents who understood and quietly admonished him to sit down until his meal arrived. Arnold would promptly obey even if he didn't understand. The chairman was only fulfilling the requirements of his position. Arnold passed away on June 26, 1981.

Beatrice continued to live alone in Torquay, She enjoyed her church, garden, quilting and neighbours. Her life was fulfilling when she could enjoy the company of others and serve them a meal. Beatrice's hamburger soup was her speciality. Everyone enjoyed her meatballs, made to perfection by following her meatball routine.

Beatrice enjoyed her coffee, a half-cup at a time to ensure a hot temperature until the last drop. Beatrice was pleased to meet granddaughter Kathy's fiancé. She was surprised to hear that Tim didn't drink coffee. She questioned, "What do you do at three o'clock in the afternoon then?"

A battered and beaten book printed in the early twenties, *The American Star Speaker and Elocutionist*, had been among Beatrice's possessions forever. Reciting poetry had been Beatrice's passion as a young maiden. She was known to entertain her youthful friends with such recitations as:

> Dark and dreary was the night,
> A storm was drawing nigh.
> In vivid streaks the lightning flashed,
> Athwart the leaden sky;
> But see from out a lonely wood,
> There steals vengeful man,
> A bloodstained club is firmly grasped,
> Within his strong right hand,
> And like a spectre from the unknown world
> He glides upon his foe.

~ A Celebration of a Century ~

> A murderous light gleams in his eye,
> As he makes ready for the blow.
> The club is raised, and then, oh—
> It falls with a sickening thud,
> And there upon the damp cold ground,
> Lays murdered—"A Potato Bug."
>
> *Anonymous*

But Beatrice's Bible was faithfully read on a daily basis. One paper tucked in the back of her Bible contained notes that she had written as she read the book of Revelation:

> What pictures of Christ do I get from this book? It is really the voice of Jesus that comes through the whole book. The angel affirms that the Word is trustworthy because it has come from God Himself. He is the one who has guided the spirit of the prophets. In the first chapter a blessing is promised to them that read aloud the Words of prophecy and blessed are those who hear and obey what is written.
>
> Christ is coming out the victor. This was written for our encouragement. Christ is the tree of life. Christ died for our sin. Christ is with us no matter what happens in our lives. Revelation is written for Christians. We should never fear the future when God lives in our heart.

Beatrice later lived with her daughter Lorna in Regina and then moved to Elmview Extended Care. She enjoyed new friends, hymn sings and the friendly atmosphere of her last home. She passed away on July 26, 1999.

Clinton and his wife Muriel moved to the farm in 1977. They had married on April 19, 1965. Clinton had farmed

The Close of a Century

from town for twelve years. Farming had always been Clinton's way of life. It was in his blood. He was the son who faithfully did the chores, worked the fields and tended to daily responsibilities so his father could be actively involved in community affairs that took him away from farm duties. It was Clinton who took the time to enjoy a cup of coffee when his mother had a newly baked cake on the table, freshly perked coffee in the pot, and it was three o'clock. Clinton was there.

Arnold had always commented that his daughter-in-law was the hardest working farm girl that he had ever met. He had praised Muriel for scrubbing out grain bins, clearing old brush from under the trees, encouraging rhubarb plants to grow, and supervising farmyard cleanup. Muriel knew how to make a garden grow. She was never idle.

Beatrice considered Muriel's discovery one day to be a miracle. Beatrice had given Arnold a gold signet ring inscribed with his initials before they were married.

Arnold lost the ring in 1928. Muriel happened to see a bright glitter in the dirt near the shop one day. She had no idea that her discovery would bring her mother-in-law such joy. Beatrice wore the ring, glittering as new, until her death. Clinton now wears the ring that was buried on the Pederson farm for close to fifty years.

The years have brought consistent change to the century farm. The towering windmill that supplied the neighbours with water during the depression had to be lowered from its near-to-the-clouds position. Clinton feared the mighty windmill might topple. No one can now quickly tell the direction of the wind with a glance. The crows have lost a lofty perch. A country landmark has hit the dust.

The granary with the upper storey came down, displacing fifteen skunks that had burrowed their home beneath it. A Quonset was built in 1979, and a heated shop in 1981.

The naturally heat-efficient chicken coop built with windows facing south had seen its best days. It, too, had to be demolished. The present owner has had to replace the farm construction that has tired from age and use. Farm machinery has to keep up with the times if a farm is to be productive. And it continues to be productive as long as the sun shines and rain falls from the wide open skies of Saskatchewan.

Clinton, too, has been involved in community organizations. He first became a Cambria #6 councillor in 1966. He joined the Estevan Co-op Board in 1985 and became a Saskatchewan Wheat Pool delegate for District One in 1995. Church participation is consistently a priority. Giving and receiving is a vital component of a satisfying life in church and district. Torquay provides such opportunities. The Pedersons have been blessed and privileged to live in such a community.

Sometimes it is necessary to go back one hundred years to realize the abundance of blessings that have been passed down through time. Homesteaders faced the winds of change as they followed their aspirations to move toward independence. Mary and Charlie Pederson's dreams of hope and opportunity for themselves and for their descendents were dreams held in trust, promoting a family continuity that has passed down for a century in the caring community of Torquay, Saskatchewan. God's strength sustained and provided for them on their journey. May the renewal of their spirit flourish in the heart of each descendent now and in future generations as they concur with the Psalmist,

"The boundary lines have fallen for me in pleasant places; I have a goodly heritage" (Psalm 16:6 NRSV).